Current Topics in Microbiology 218 and Immunology

Springer
Berlin
Heidelberg
New York
Barcelona
Budapest
Hong Kong
London
Milan
Paris
Santa Clara
Singapore
Tokyo

Adeno-Associated Virus (AAV) Vectors in Gene Therapy

Edited by K.I. Berns and C. Giraud

With 38 Figures

Springer

Professor KENNETH I. BERNS, M.D., Ph.D.
CATHERINE GIRAUD, Ph.D.

Cornell University Medical College
Hearst Microbiology Research Center
Department of Microbiology
1300 York Avenue, Box 62
New York, NY 10021
USA

Cover illustration: The inverted terminal repeat of the adeno-associated virus genome is shown in the folded conformation which maximizes base pairing, The inverted terminal repeat plays a significant role in the regulation of viral gene expression, DNA replication, and integration into and excision from the integration site in the human genome. The red letters indicate the binding site of the viral regulatory protein "rep 68/78." "trs" indicates the site nicked by the rep 68/78 protein.

Cover design: Design & Production, Heidelberg

ISSN 0070-217X
ISBN-13: 978-3-642-80209-6 e-ISBN-13: 978-3-642-80207-2
DOI: 10.1007/978-3-642-80207-2

Softcover reprint of the hardcover 1st edition 1996
Library of Congress Catalog Card Number 15 - 12910

Typesetting: Scientific Publishing Services (P) Ltd, Madras

SPIN: 10528856 27/3020/SPS – 5 4 3 2 1 0 – Printed on acid-free paper

Preface

Human gene therapy holds great promise for the cure of many genetic diseases. In order to achieve such a cure there are two requirements. First, the affected gene must be cloned, its sequence determined and its regulation adequately characterized. Second, a suitable vector for the delivery of a good copy of the affected gene must be available. For a vector to be of use several attributes are highly desirable: these include ability to carry the intact gene (although this may be either the genomic or the cDNA form) in a stable form, ability to introduce the gene into the desired cell type, ability to express the introduced gene in an appropriately regulated manner for an extended period of time, and a lack of toxicity for the recipient. Also of concern is the frequency of cell transformation and, in some cases, the ability to introduce the gene into nondividing stem cells. Several animal viruses have been tested as potential vectors, but none has proven to have all the desired properties described above. For example, retroviruses are difficult to propagate in sufficient titers, do not integrate into nondividing cells, and are of concern because of their oncogenic properties in some hosts and because they integrate at many sites in the genome and, thus, are potentially insertional mutagens. Additionally, genes introduced by retroviral vectors are frequently expressed for relatively short periods of time. A second virus used as a vector in model systems has been adenovirus (Ad). The major deficit is that Ad does not integrate and functions as a transient vector, necessitating repeated administration with the possibility of a toxic immune response. A third virus of interest as a possible vector is adeno-associated virus (AAV), a small DNA virus which has many of the properties desired in a vector. This volume is devoted to descriptions of the manner in which AAV has been used as a vector for gene therapy in cell culture and in model animal systems. The initial chapter is a general overview of the biology and molecular biology of AAV with an emphasis on those properties directly pertinent to use as a vector. Also included are discussions of some of the questions that remain to be answered before the possibility of clinical application. Subsequent chapters describe the preparation of AAV vectors, including attempts to develop packaging cell lines. The remaining chapters describe the construction and use in vitro of AAV virion vectors for the potential treatment of diseases of the hematopoietic system, cystic fibrosis, and AIDS. Two fundamental questions remain to be resolved. The first is practical: how to produce vectors in sufficient quantity for clinical use. The second is both theoretical and practical: should AAV vectors be designed to integrate at a specific site in the human genome. The reader should consider both of these issues as the chapters on the use of AAV as a vector are read.

July 1996 KENNETH I. BERNS

List of Contents

List of Contributors

(Their addresses can be found at the beginning of their respective chapters.)

Biology of Adeno-associated Virus

K.I. Berns and C. Giraud

1 Adeno-associated Virus Life Cycle

Adeno-associated Virus (AAV) is classified as a member of the family Parvoviridae (Siegl et al. 1985; Berns 1990a). Members are small, nonenveloped, icosahedral viruses (diameter ca. 20–26 nm) with linear, single-stranded DNA genomes of 4.7–6 kb. Parvoviridae have been isolated from many species ranging from insects to humans. AAV is assigned to the genus *Dependovirus*, so named because the virus was discovered as a contaminant in purified adenovirus (Ad) stocks and in most instances does not productively infect cells in culture unless there is a coinfection by an unrelated helper virus, which is most commonly Ad, but also can be any type of herpesvirus (Atchinson et al. 1965; Hoggan et al. 1966; Buller et al. 1981). Various serotypes have been isolated from birds and many mammalian species, including humans. About 90% of U.S. adults are seropositive, but in no case has the virus been implicated as the etiological agent for a human disease or as the cause of disease in any other species. Because of the requirement for a helper coinfection for productive infection in cell culture, AAV was long considered to be a defective virus. Detailed studies described below have demonstrated that the virus is not defective, but rather preferentially establishes a latent infection in a healthy cell and is only induced to undergo productive vegetative multiplication when the host cell is stressed.

Department of Microbiology, Hearst Microbiology Research Center, Cornell University Medical College, 1300 York Avenue, New York, NY 10021, USA

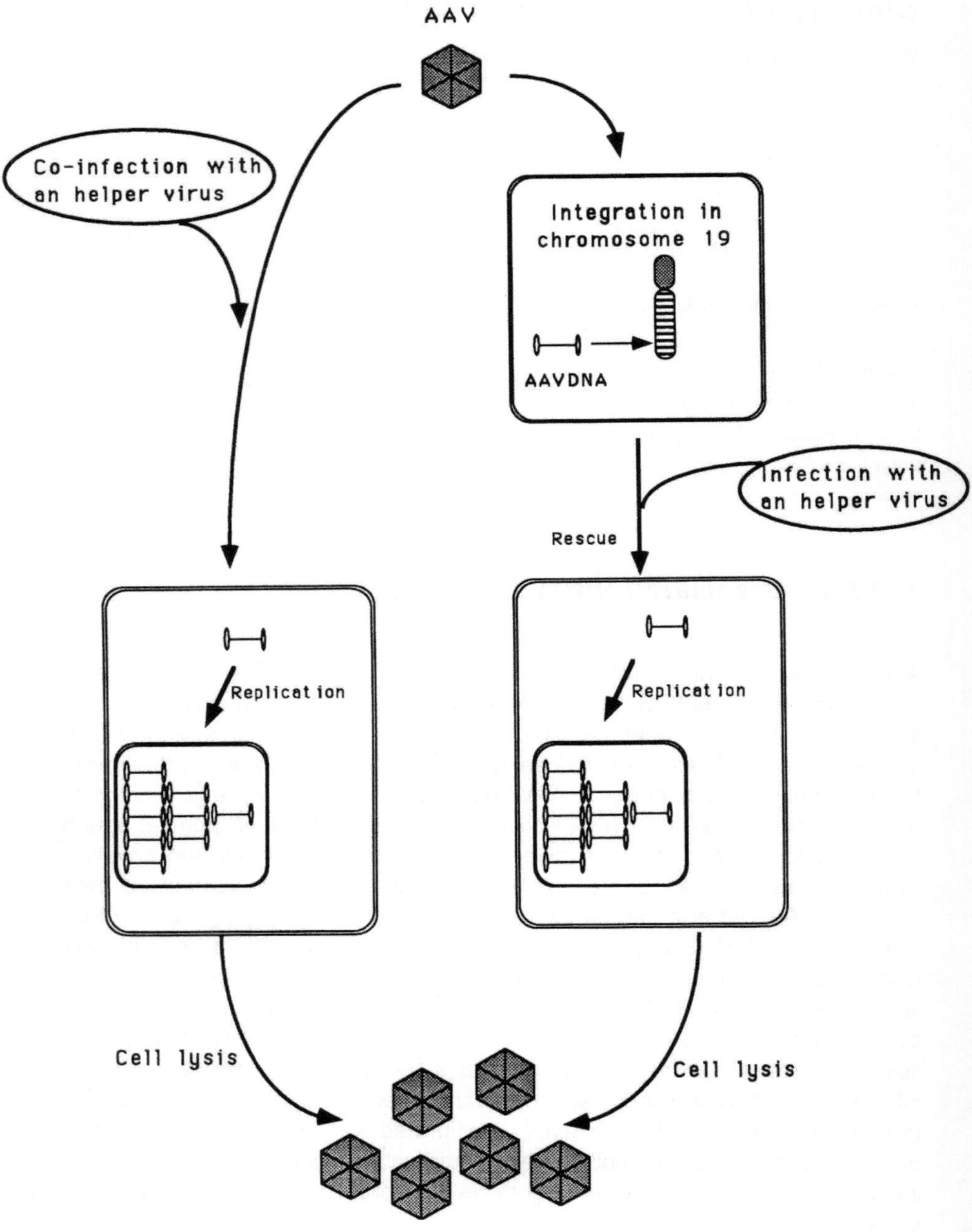

Fig. 1. Adeno-associated virus (AAV) life cycle. Under nonpermissive conditions AAV integrates into the q arm of human chromosome 19 where it remains silent until challenged by a helper virus, e.g., adenovirus. This leads to rescue of the integrated virus from the chromosome and induction of the lytic cycle. Under permissive conditions, i.e., in the presence of a helper virus, such as adenovirus, AAV replicates, resulting in host cell lysis

There are two distinct intracellular phases to the AAV life cycle (Fig. 1; for review see LEONARD and BERNS 1994b; BERNS and LINDEN 1995). In the absence of coinfection by a helper virus, the AAV virion is able to enter the cell and the genome is uncoated. There is a limited expression of the AAV regulatory proteins; the effect of the limited expression is to repress any further viral gene expression and to inhibit measurable amounts of viral DNA replication. The consequence of the negative effects on viral macromolecular synthesis appears to be enhancement of the integration of the viral genome into the genome of the host cell and to thus establish a latent infection (HOGGAN et al. 1972; HANDA et al. 1977; CHEUNG et al. 1980). Any DNA introduced into a mammalian cell by any technique (e.g., microinjection, transfection, or infection) can recombine with cellular genomic DNA, almost always by nonhomologous recombination. AAV is unique in that it is the first example of an exogenous DNA which has been demonstrated to preferentially integrate at a specific site in the human genome, on the q arm of chromosome 19 (KOTIN et al. 1990, 1991, 1992; SAMULSKI et al. 1991; SAMULSKI 1993). Studies of cells in culture latently infected by AAV have demonstrated that there are effects on the cellular phenotype which are usually subtle and affect the ability of the cell to respond to stress (YALKINOGLU et al. 1988; YAKOBSON et al. 1987, 1989; BANTEL-SCHAAL and STÖHR 1992; BANTEL-SCHAAL 1991; WINOCOUR et al. 1988, 1992; WALZ and SCHLEHOFER 1992; WALZ et al. 1992; KLEIN-BAUERNSCHMITT et al. 1992).

The second intracellular phase of the AAV life cycle is productive infection, which most often occurs in the presence of infection by a helper virus (either Ad or a herpesvirus) (ATCHINSON et al. 1965; HOGGAN et al. 1966; BULLER et al. 1981). In the case of Ad, infection can either precede that of AAV, occur at the same time, or involve superinfection of a cell latently infected by AAV. Much the same is true when the helper is a herpesvirus, but the rapid lytic nature of some herpesvirus infections means that infection by the helper cannot precede that of AAV by a significant amount. For both Ad and herpes simplex virus (HSV) the genes involved in the helper effect have been determined. All of the Ad early genes which are involved in regulation of gene expression (El a and b, E2a, E4, and VA RNAs) provide helper functions (RICHARDSON and WESTPHAL 1981, 1984; RICHARDSON et al. 1980; CHANG et al. 1989; JAY et al. 1979; JANIK et al. 1989; WEST et al. 1987; SAMULSKI and SHENK 1988). Interestingly, the E2b gene products (DNA polymerase and terminal protein) which are directly involved in Ad DNA replication are not required (STILLMAN et al. 1981, 1982). The situation is somewhat different in the case of HSV 1 helper function. Not only are genes involved in gene regulation (e.g., ICP O and ICP 4) needed, but also the viral DNA polymerase and possibly the viral helicase (MISHRA and ROSE 1990; WEINDLER and HEILBRONN 1991). This difference in helper functions may reflect differential effects on host DNA synthesis machinery during Ad and HSV infection. The common denominator is the involvement of genes affecting regulation of gene expression. These genes affect cellular gene expression as well as viral expression. Thus, the notion has developed that the generic role of the helper virus is to alter the intracellular milieu to allow the expression of AAV genes which are required for the production for progeny virions.

The current working model has the following elements: (1) AAV is a virus with a life cycle in which latent infection is a major component. (2) By virtue of a latent infection the viral genome is perpetuated by the cell. (3) As long as the host cell is healthy, AAV gene expression is repressed and the latent state is maintained. (4) If the cell is stressed, the intracellular milieu is altered so that stress response genes are expressed. (5) The regulatory state which permits the expression of cellular response genes also permits expression of the AAV genes which are required for viral replication. (6) Progeny virions are produced which are released to infect new, healthy cells in order to reestablish the latent state. There are several lines of evidence which support the model. Ad Ela activates certain cell genes which encode heat shock proteins (NEVINS 1982; WU et al. 1986). Conversely, heat shock of human cells renders them permissive for the expression of AAV genes. Treatment of certain types of mammalian cells in culture with genotoxic agents (e.g., UV irradiation, γ-irradiation, various chemical carcinogens, or certain metabolic inhibitors) renders the treated cells permissive for productive AAV infection in the absence of helper virus (SCHLEHOFER et al. 1986; YAKOBSON et al. 1987; YALKINOGLU et al. 1988). Although the viral yield per cell is several logs greater in the presence of a helper virus, the above experiments are strong indicators that AAV is not a defective virus, but rather a virus strongly oriented to latency. In many ways it seems the animal virus equivalent of λ bacteriophage.

2 Genetics

The AAV2 genome has been sequenced. Within the 4680 nucleotide (nt) sequence there are two large open reading frames (orfs), which occupy the right and left halves of the genome, respectively (SRIVASTAVA et al. 1983). The right side orf encodes three structural proteins (VP 1–3) with overlapping amino acid sequences. The three proteins from a common sequence are the consequence of differential splicing and the use of an unusual initiator codon in one case (Fig. 2; CASSINOTTI et al. 1988; BECERRA et al. 1988; TREMPE and CARTER 1988; MURALIDHAR et al. 1994). The left side orf encodes four regulatory proteins, again with overlapping amino acid sequences (MENDELSON et al. 1986). Because frame shift mutations at almost any point in the orf inhibit DNA replication, the orf has been dubbed the *rep* gene and the four proteins encoded have been designated on the basis of perceived molecular weight as Rep 78, 68, 52, and 40 (HERMONAT et al. 1984; TRATSCHIN et al. 1984a; SENAPATHY et al. 1984). Promoters have been mapped using a scale of 100 map units at map positions (mps) 5, 19, and 40. Rep 78 and 68 are translated from the unspliced and spliced forms, respectively, from the p5 transcript driven by the promoter at mp5. Similarly, Rep 52 and 40

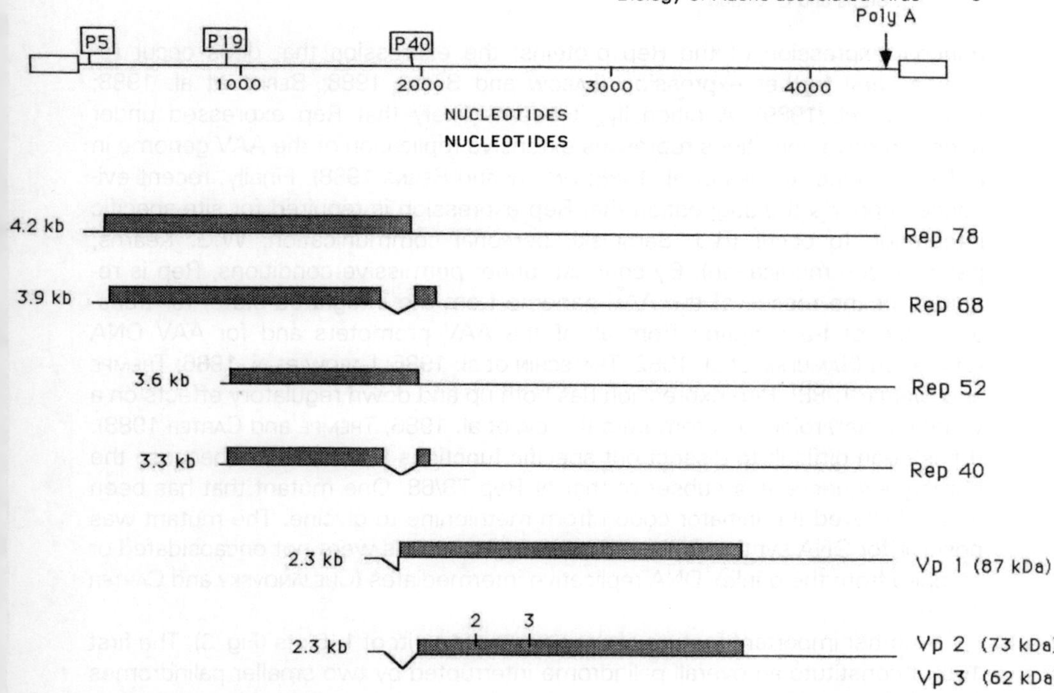

Fig. 2. Genetic map of adeno-associated virus (AAV) genome. Several transcripts are initiated from three promoters at map positions 5, 19 and 40 (e.g., p5, p19 and p40) and terminate at the same polyadenylation site at map position 95. The presence of an intron generates both spliced and unspliced forms of the mRNAs. Two transcripts, unspliced and spliced, initiated from the p5 promoter encode for the two large rep proteins, respectively, Rep 78 and Rep 68. The p19 promoter initiates an unspliced and spliced transcript encoding the smallest rep proteins, respectively, Rep 52 and 40; p40 initiates all three capsid mRNAs. VP2 and VP3 are produced from the same spliced transcript. VP2 uses an alternative ACG codon; VP1 is produced by an alternative splicing event

are translated from the p19 promoter and VP 1–3 are translated from alternatively spliced forms of the p40 promoter. (VP 2 initiates at an ACG codon upstream from the AUG used for VP 3).

Rep 78 and 68 have been extensively characterized in terms of their properties and roles in the viral life cycle. To date, significant differences in their properties have not been detected. Both proteins bind to a specific sequence (5′-GCTCGCTCGCTC–3′) found in the inverted terminal repeat (itr) (McCarty et al. 1994b; Chiorini et al. 1994). Similar sequences are found upstream of all three AAV promoters (McCarty et al. 1994a). Binding is enhanced when the itr is folded on itself (McCarty et al. 1994b). Rep 78/68 functions as both an ATPase and a helicase (Im and Muzyczka 1989, 1990, 1992). When Rep 78/68 binds to the itr, it functions as a site-specific nickase which cleaves after nt 124 (Snyder et al. 1993). The sequence recognized is 5′–T/TGG. Finally, Rep 78/68 forms homo-oligomers. The two proteins are required for every phase of the AAV life cycle. Under nonpermissive conditions (neither helper nor stress) there is minimal expression of the Rep proteins; the expression that does occur represses any further expression (LABOW and BERNS 1988; BERNS et al. 1988; BEATON et al. 1989). Additionally, it seems likely that Rep expressed under nonpermissive conditions represses extensive replication of the AAV genome in a direct manner (BERNS et al. 1988; LABOW and BERNS 1988). Finally, recent evidence supports the suggestion that Rep expression is required for site-specific integration to occur (R.J. Samulski, personal communication; W.G. Kearns, personal communication). By contrast, under permissive conditions, Rep is required for the rescue of the AAV genome from the integrated state, for transactivation of transcription from all of the AAV promoters and for AAV DNA replication (SAMULSKI et al. 1982; TRATSCHIN et al. 1986; LABOW et al. 1986; TREMPE and CARTER 1988). Rep expression has both up and down regulatory effects on a variety of heterologous promoters (LABOW et al. 1986; TREMPE and CARTER 1988). It has been difficult to dissect out specific functions for Rep 52/40 because the primary sequence is a subset of that of Rep 78/68. One mutant that has been created altered the initiator codon from methionine to glycine. The mutant was positive for DNA synthesis, but mature single strands were not encapsidated or stripped from the duplex DNA replicative intermediates (CHEJANOVSKY and CARTER 1989).

Another important feature of the genome is an itr of 145 nts (Fig. 3). The first 125 nt constitute an overall palindrome interrupted by two smaller palindromes (nts 42–62 and 64–84), one on either side of the overall axis of symmetry. When folded on itself, the itr forms a t-shaped structure which is thought to be important in the replication of the virus (SRIVASTAVA 1987; BOHENZKY et al. 1988). The itr has been demonstrated to constitute an important cis-acting signal which functions as an enhancer, is critical for the negative regulation of DNA replication under nonpermissive conditions, serves as the origin of DNA replication (ori) and the primer for initiation of DNA synthesis under permissive conditions, is required for the site-specific integration of AAV and is essential for rescue of the viral genome from the integrated state (BEATON et al. 1989; LABOW et al. 1986; LABOW and BERNS 1988; SAMULSKI et al. 1983, 1989; TRATSCHIN et al. 1984a; SENAPATHY et al. 1984).

3 Latent Infection

The facility with which AAV can establish a latent infection is the primary factor in its potential utility as a vector for human gene therapy (SAMULSKI 1993). The occurrence of natural latent infection by AAV was discovered by screening of primary primate cell cultures which were to be used for vaccine production. In the original study it was found that up to 20% of lots of primary African green monkey kidney cells and 1%–2% of human embryonic kidney cell lots would release AAV when challenged by superinfection with Ad (HOGGAN et al. 1972).

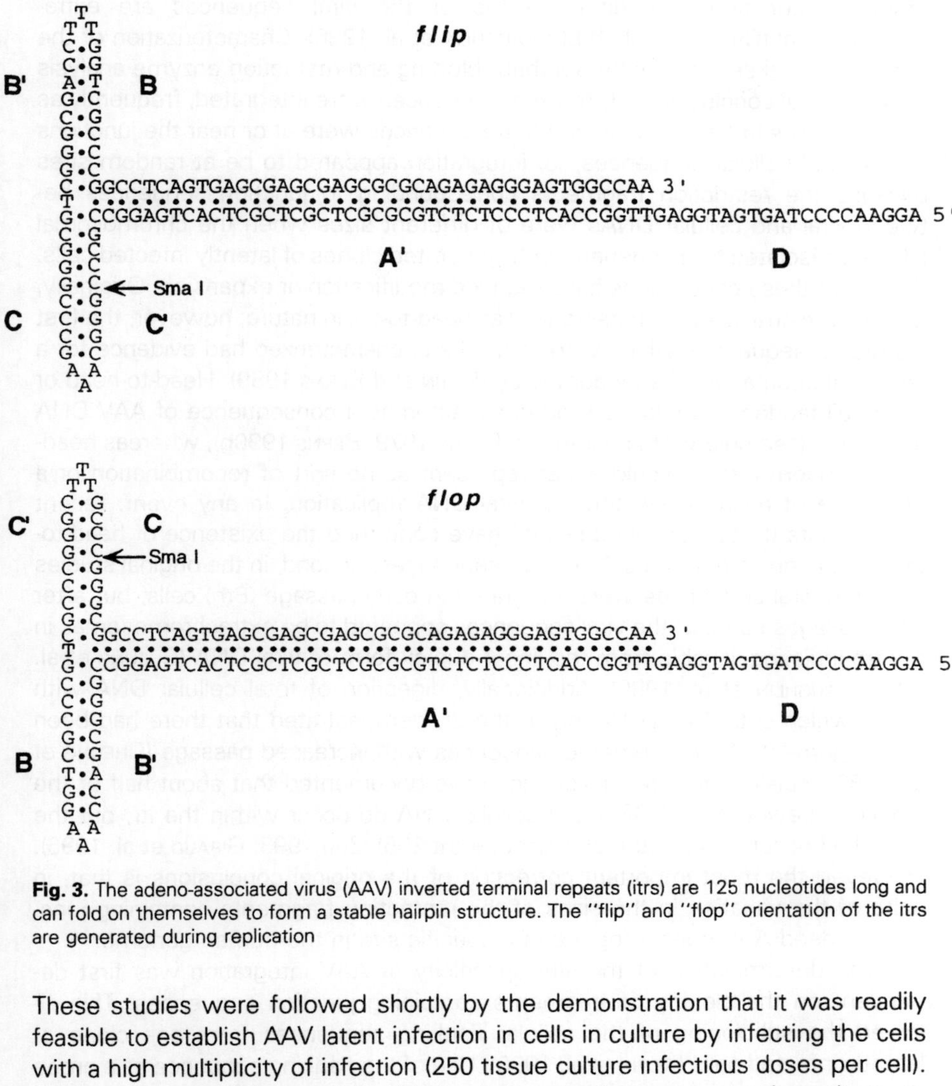

Fig. 3. The adeno-associated virus (AAV) inverted terminal repeats (itrs) are 125 nucleotides long and can fold on themselves to form a stable hairpin structure. The "flip" and "flop" orientation of the itrs are generated during replication

These studies were followed shortly by the demonstration that it was readily feasible to establish AAV latent infection in cells in culture by infecting the cells with a high multiplicity of infection (250 tissue culture infectious doses per cell). Under these conditions, which involved no selection pressure, the culture remained positive (able to release AAV only when challenged by Ad infection) for >40 passages. Cell clones isolated at that point remained positive for >150 passages (Berns et al. 1975). The existence of latently infected cells in culture has allowed detailed characterization at the molecular level of the AAV DNA sequences in the latent state (Handa et al. 1977; Cheung et al. 1980; Laughlin et al. 1986; McLaughlin et al. 1988; Kotin et al. 1991; Samulski et al. 1989, 1991; Goodman et al. 1994).

Initial studies indicated that the AAV sequences were integrated into the cellular genome during latency (Handa et al. 1977; Cheung et al. 1980; Laughlin et al. 1986). More recent studies have supported the original conclusions, although under some conditions some of the viral sequences are extrachromosomal (CHEUNG et al. 1980; LAUGHLIN et al. 1986). Characterization of the integrated viral sequences by Southern blotting and restriction enzyme analysis led to several conclusions: (1) the viral sequences were integrated, frequently as several copies in a tandem array; (2) itr sequences were at or near the junctions of viral and cellular sequences; (3) integration appeared to be at random sites because the restriction fragments corresponding to junction sequences between viral and cellular DNAs were of different sizes when the chromosomal DNA was isolated from independently generated clones of latently infected cells. Several of these conclusions have required modification or expansion. Originally, the tandem arrays were characterized as head-to-tail in nature; however, the first integrated sequences which were cloned and characterized had evidence for a tail-to-tail orientation of a tandem array (KOTIN and BERNS 1989). Head-to-head or tail-to-tail tandems would have been expected as a consequence of AAV DNA replication (see below, HAUSWIRTH and BERNS 1979; BERNS 1990b), whereas head-to-tail tandem arrays would either represent some sort of recombination or a new type of rolling circle form of viral DNA replication. In any event, recent experiments (to be described below) have confirmed the existence of head-to-tail orientation of integrated AAV DNA sequences. Second, in the original studies all of the viral sequences were integrated in early passage (8th) cells, but after 118 passages some of the viral sequences appeared to be extrachromosomal, in a form indistinguishable from the linear duplex form of viral DNA (CHEUNG et al. 1980; LAUGHLIN et al. 1986). Additionally, digestion of total cellular DNA with *SmaI*, which cuts AAV DNA only in the itr, demonstrated that there had been rearrangement of the integrated sequences with increased passage (CHEUNG et al. 1980). Finally, more recent studies have documented that about half of the junctions between viral DNA and cellular DNA do occur within the itr, but the other half occur close to the p5 promoter (nt 255) (ZHU 1993; GIRAUD et al. 1995). However, the most important correction of the original conclusions is that, in spite of the variation in the sizes of the restriction fragments containing junctions, indeed AAV does integrate at a specific site in the human genome.

The determination of the site specificity of AAV integration was first dependent on the cloning of fragments containing junction sequences. This allowed the subcloning of the cellular flanking sequences from one clone of latently infected cells (KOTIN et al. 1990). Use of the flanking sequence as a probe demonstrated that the sequence was single copy in the uninfected cell. Extra fragments hybridized to the probe when genomic DNA from the latently infected cell line was tested. The assumption was that the extra fragments represented cellular sequences disrupted by integration of the viral DNA (KOTIN et al. 1990). If this was true and AAV integrated at a specific site, comparable extra bands should have been observed when genomic DNAs from independently derived clones of latently infected cells were tested. The results were as predicted in the majority of cell clone DNAs tested (67%); thus, it was concluded that AAV does integrate in a site-specific manner in continuous lines of human cells (KOTIN et al. 1990, 1991). The integration site was mapped to human chromosome 19, initially by Southern blot hybridization to genomic DNAs from mouse-human hybrid cell lines. Subsequently, both the site specificity of integration and the localization of the integration site to the q arm of chromosome 19 have been confirmed by several laboratories using fluorescent in situ hybridization (KOTIN et al. 1991; SAMULSKI et al. 1991). It should be noted that not every integration event occurs at the specific site, although a majority do (WALZ and SCHLEHOFER 1992). More importantly, as documented below, even at the specific site, the actual junctions between viral and cellular sequences vary by several hundred nucleotides (SAMULSKI et al. 1991; KOTIN et al. 1992; ZHU 1993; GIRAUD et al. 1995; GOODMAN et al. 1994).

A significant question with regard to understanding the basic mechanisms underlying site-specific integration by AAV is whether the specific site is determined by the DNA sequence or by a higher order chromatin structure present at the specific site in chromosome 19 (q13.3-qter). An 8.2 kb fragment (aavS1 sequence) encompassing the specific integration site has been cloned and the 5′ 4 kb have been sequenced (KOTIN et al. 1992). The sequence has a number of interesting features: (1) an orf which is transcribed at low levels in several tissues; (2) a CpG island upstream from the orf; (3) a tandem array of ten copies of a 35 nt minisatellite sequence (a sequence which occurs at 60 sites in the human genome, all on the q arm of chromosome 19; DAS et al. 1987); (4) a higher than expected number of direct repeats of dodecanucleotide sequences in the regions on either side of the orf. The overall GC content is 65%, but the GC content of the 5′ 1100 nt is 82%. However, none of these features would appear to confer the required specificity to account for a single site of integration in the human genome. To address this question in more detail an additional experimental approach has been developed.

In order to assess whether the primary DNA sequence was the determining factor for the specificity of integration site, the cloned integration site was placed at an additional locus in the cell (GIRAUD et al. 1994). In this way it was considered that any higher order chromatin structure effects peculiar to the q arm of chromosome 19 could be factored out. For this purpose a shuttle vector derived from Epstein-Barr virus (EBV) was used (for review see MARGOLSKEE 1992). The vector had the properties of being able to be propagated extrachromosomally in both bacteria and mammalian cells and carried selectable markers for both hosts. Human cell lines were established which carried ca. 50–150 copies of the vector into which had been inserted the 8.2 kb DNA fragment containing the specific site of integration. The cell lines were then infected with wild-type AAV. At 48 h post-infection, a Hirt extraction procedure was performed to isolate low molecular weight DNA, and *E. coli* were transformed with the DNA in the Hirt extract. Bacterial colonies which grew up under the conditions of selection were screened with an AAV probe by DNA hybridization to see if any of the colonies contained a plasmid into which AAV had integrated. Approximately 1%–2% of such colonies were positive for AAV sequences when the shuttle vector contained the 8.2 kb fragment representing the specific site of integration. By contrast, no positive colonies were detected when vectors containing control

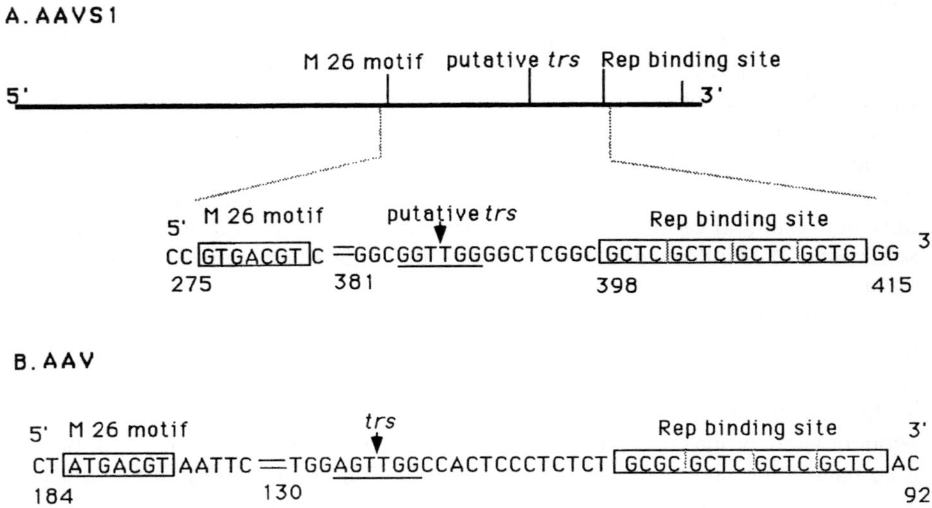

Fig. 4A,B. Comparison of the adeno-associated virus (AAV) S1 510 nucleotides (nts) and AAV itr sequences. **A** Characteristics of the 510 nts at the 5′ end of the AAVS1 preintegration sequence. The 510 nts are part of a CpG island (nts 1–900) and a region with a high frequency of short repeats. The enlarged part of the sequence shows the rep binding site, a putative terminal resolution site (*trs*) and the M26 motif from yeast. **B** AAV2 viral DNA sequence. The nucleotide numbers indicated refer to the complementary strand of the AAV sequence. The trs is indicated by a *vertical arrow*

inserts (random human DNA sequences) were tested. To determine the minimal segment required to detect AAV integration, serial deletions were performed. In summary, the minimal sequence required was contained within the 5′ terminal 510 nts of the original 8.2 kb fragment. Conversely, no fragment from which the 5′ terminal 510 nts were deleted led to integration by AAV.

Thus, the critical sequences for recognition of a specific site of integration were within a rather restricted segment of the originally defined site. Inspection of the sequence revealed three short sequences which are candidates to be important *cis*-active sequences (Fig. 4). The first is a sequence which corresponds to a binding site for Rep 78/68 (Weitzman et al. 1994; Chiorini et al. 1994). Indeed, Rep can bind to this sequence. The second is a sequence which matches the terminal resolution site trs-specific nick site for Rep 78/68 (Snyder et al. 1993). It also appears that this sequence is cleaved in this context when Rep binds to the fragment (Urcelay et al. 1995). The spacing between the Rep binding site and the trs in the integration site are quite comparable to the spacing in the AAV itr. A subfragment of the specific site of integration carrying these two motifs can undergo asymetric replication in vitro (Urcelay et al. 1995). The third short sequence which may be of importance is a heptamer which corresponds to a sequence described in bacteriophage and in yeast (M26) to enhance recombination (Schuchert et al. 1991; Ponticelli and Smith 1992). Site-specific mutagenesis experiments are in progress to determine whether, in fact, these three sequences are both necessary and sufficient to direct integration. An important point to note is that the actual junctions between AAV and cellular sequences that have been identified lie 3′ to the 510 nt sequence (KOTIN et al. 1992; SAMULSKI et al. 1991; ZHU 1993; GOODMAN et al. 1994). Using the EBV shuttle vectors, the majority of the junctions were mapped near the rep binding site in the AAVS1 sequence. The remainder mapped 3′ to the 510 nt minimal sequence (GIRAUD et al. 1995). If the conclusions reached from the experiments described above are correct, they would imply that the signals identified direct site-specific integration but that the recombination events could occur at some distal point.

The recombinants formed have been characterized with respect to the AAV sequences present, their orientation, and the junctions with vector DNA (GIRAUD et al. 1995). Some 20% (9/43) of the recombinants isolated apparently contained the entire AAV sequence. AAV protein expression could be detected from some of the recombinants transfected in Ad-infected cell lines. Two of the recombinants tested expressed the AAV protein at a level similar to the one seen with the pSM620 plasmid (SAMULSKI et al. 1982). These two plasmids were also able to be rescued, to replicate and to produce infectious virus particles. The key factor in the ability to be rescued appeared to be the presence of an itr at one end which was extended by the addition of 25 nts complementary to nts 126–145 of the normal itr. The consequence of the extension was to create an extended palindromic sequence of the type normally found between two copies of the genome when present as a head-to-tail tandem (Fig. 5). Most of the remaining recombinants had deletions in the left half, expanded in some cases to the right half, of the genome. When the disruption was at or near the p5 promoter the deleted genomes were present in a type of head-tail tandem orientation (Fig. 5). This was comparable to the orientations observed in chromosomal integration (CHEUNG et al. 1980; LAUGHLIN et al. 1986). The recombinants derived from an EBV shuttle vector carrying the 5′ terminal 510 bp aavS1 sequence present, so far, a different AAV sequence organization. They appear to contain AAV sequences either from the right part or the left part of the genome. In the case of each recombinant, one of the junctions with vector DNA was within the aavS1 sequence present in the construct. The other junction was in

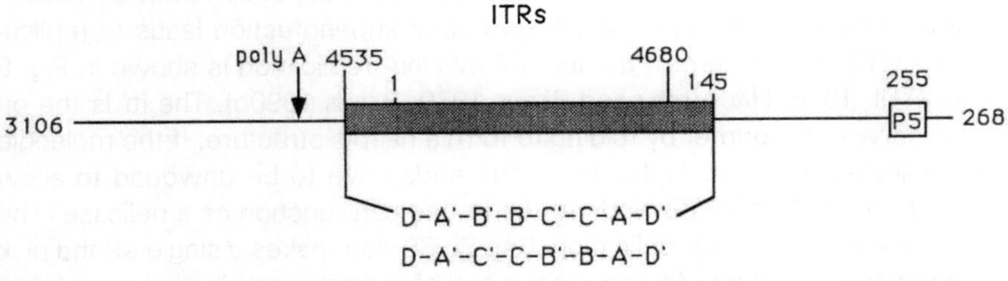

Fig. 5. The adeno-associated virus (AAV) sequences, inserted in the aavS1 locus carried by an Epstein-Barr Virus episomal vector, are found in a head-to-tail conformation with a "flip" or "flop" orientation of the inverted terminal repeat (itr)

non-aavS1 sequence. About half of the junctions with aavS1 occurred within the itr; the other half occurred at or near the p5 promoter, again as has been observed with chromosomal integration. The conclusion from these experiments is that the model system is a good approximation of the natural integration event and will allow determination of the key signals involved in site-specific integration. To the extent that site-specific integration is determined to be a desirable characteristic of a vector for gene therapy, an understanding of the mechanisms involved is of signficance.

4 Rescue and Replicative Infection

The study of AAV genetics has been greatly facilitated because the cloned duplex form of the DNA in a pBR322 vector is infectious (Samulski et al. 1982; Laughlin et al. 1983). When transfected into a human cell infected by Ad, the AAV DNA is rescued from the vector backbone and replicated and progeny virions are produced. Thus, any specific mutation can be introduced into the AAV genome and propagated in *E. coli* and the phenotype of the mutant assessed directly (Samulski et al. 1983; Hermonat et al. 1984; Senapathy et al. 1984; Tratschin et al. 1984a). The rescue of the viral genome from the vector would appear to be a model for rescue of the genome from the integrated state in the latently infected cell. Several conclusions have been drawn from this type of experiment: (1) The itr is required for rescue. (2) Deletions within one itr can be repaired if the other itr is intact or missing no more than 20–25 nts. (3) A deletion at one end that extends beyond the itr blocks rescue, presumably, because there is no template to copy for repair at the other terminal. (4) Rescue normally requires an intact, functional *rep* gene and an appropriate intracellular milieu to allow expression of the gene; usually, this involves superinfection by a helper Ad or herpesvirus. However, it should be noted that after many passages(>100) of clones of latently infected cells there is evidence of a low level of spontaneous rescue of the integrated AAV genome. The rescued form was characterized as a linear duplex molecule with a sequence orientation indistinguishable from the double-stranded form of virion DNA (Cheung et al. 1980; Laughlin et al. 1986).

Rescue as a consequence of a helper virus superinfection leads to replication of AAV DNA. The current model of AAV DNA replication is shown in Fig. 6 (Straus et al. 1976; Hauswirth and Berns 1979; Berns 1990b). The itr is the ori and also serves as a primer by folding to form a hairpin structure. If the molecule to be replicated is in the duplex form, the ends have to be unwound to allow hairpin formation. Rep 78/68 binds to the itr and can function as a helicase. The protein is required for DNA replication. Rep 78/68 also makes a single-strand nick at a sequence-specific site (the trs) at the end of the palindromic sequence (after nt 124); this seems likely to play a role in rescue of the viral genome from the integrated state and is important to allow restoration of the full length extended

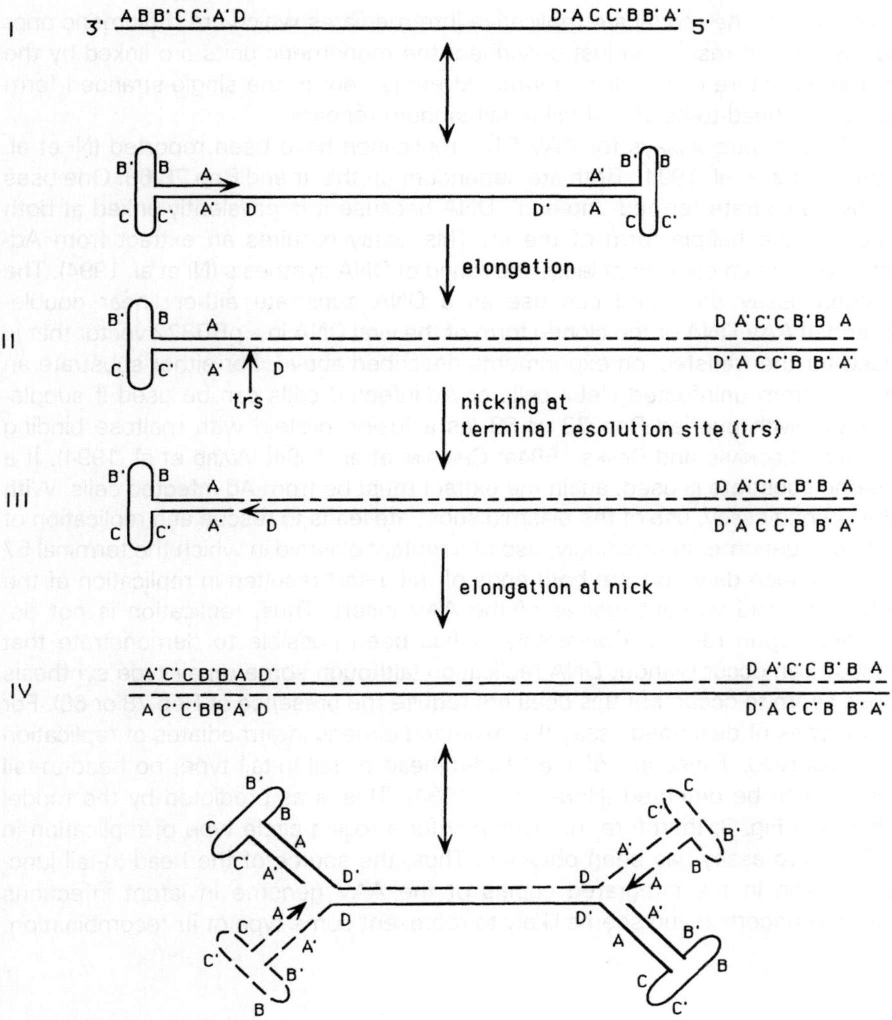

Fig. 6. Model for adeno-associated virus (AAV) replication. The folding of the inverted terminal repeat (itr) initiates the elongation from the 3′ end of the molecule (*I*) to generate a replicative form intermediate (*II*). A nick at the terminal resolution site (*trs*) is required for the resolution of the duplex form and reinitiation of DNA replication (*III, IV*)

form of the terminal of the parental strand with concomitant production of the equivalent extended form of the progeny strand. Because the primer is a DNA hairpin, which covalently links the old and new strands, the nicking activity of Rep 78/68 creates a nick with a 3′ OH which can act as a primer for synthesis of a sequence complementary to the displaced hairpin sequence. In this way the terminal hairpin sequence is transferred from the 3′ end of the parental strand to the 5′ end of the progeny strand and is inverted in the process. An important prediction of the model are replicative intermediates which are multimeric prior to the type of resolution just described; the monomeric units are linked by the hairpin structure of the itr and when stretched out in the single-stranded form represent head-to-head and tail-to-tail tandem repeats.

Two in vitro assays for AAV DNA replication have been reported (Ni et al. 1994; Ward et al. 1994). Both are dependent on the itr and Rep 78/68. One uses a DNA substrate termed "no-end" DNA because it is covalently linked at both ends by the hairpin form of the itr. This assay requires an extract from Ad-infected human cells for at least one round of DNA synthesis (Ni et al. 1994). The second assay described can use as a DNA substrate either linear double-stranded AAV DNA or the cloned form of the viral DNA in a pBR322 vector that is used for the transfection experiments described above. For either substrate an extract from uninfected HeLa cells or Ad-infected cells can be used if supplemented with purified Rep 78 or 68 (as a fusion protein with maltose binding protein) (Leonard and Berns 1994a; Chiorini et al. 1994; Ward et al. 1994). If a no-end substrate is used, again the extract must be from Ad-infected cells. With the second assay, use of the plasmid substrate leads to rescue and replication of the AAV genome. Interestingly, use of a mutant plasmid in which the terminal 57 nts had been deleted from both ends of the insert resulted in replication of the intact plasmid without rescue of the AAV insert. Thus, replication is not dependent upon rescue. Conversely, it has been possible to demonstrate that rescue can occur without DNA replication (although some repair type synthesis does seem to occur, but this does not require the presence of Rep 78 or 68). For both types of described assay the expected dimeric intermediates of replication are observed. These are of the head-to-head or tail-to-tail type; no head-to-tail dimers can be detected (Hong et al. 1994). This is as predicted by the model shown in Fig. 3; therefore, no evidence for a rolling circle type of replication in the in vitro assay has been obtained. Thus, the source of the head-to-tail junctions seen in the integrated copies of the AAV genome in latent infections remains uncertain and seems likely to represent some type of itr recombination.

5 Adeno-associated Virus Vectors for Gene Therapy

Based on some of its properties discussed above, such as non-pathogenicity, broad host range of infectivity, including nondividing cells, and preferential integration in chromosome 19q, AAV virus has recently become an attractive tool for gene therapy (Carter 1992; Muzyczka 1992; Xiao et al. 1993; Samulski 1993; Kotin 1994).

AAV recombinant viruses carrying a gene of interest to be delivered to the cells have been prepared as shown in Fig. 7. Cells are infected by an adenovirus helper and cotransfected by two vectors: one carries the gene of interest inserted between the two itrs of the AAV genome and the other carries the *cap*

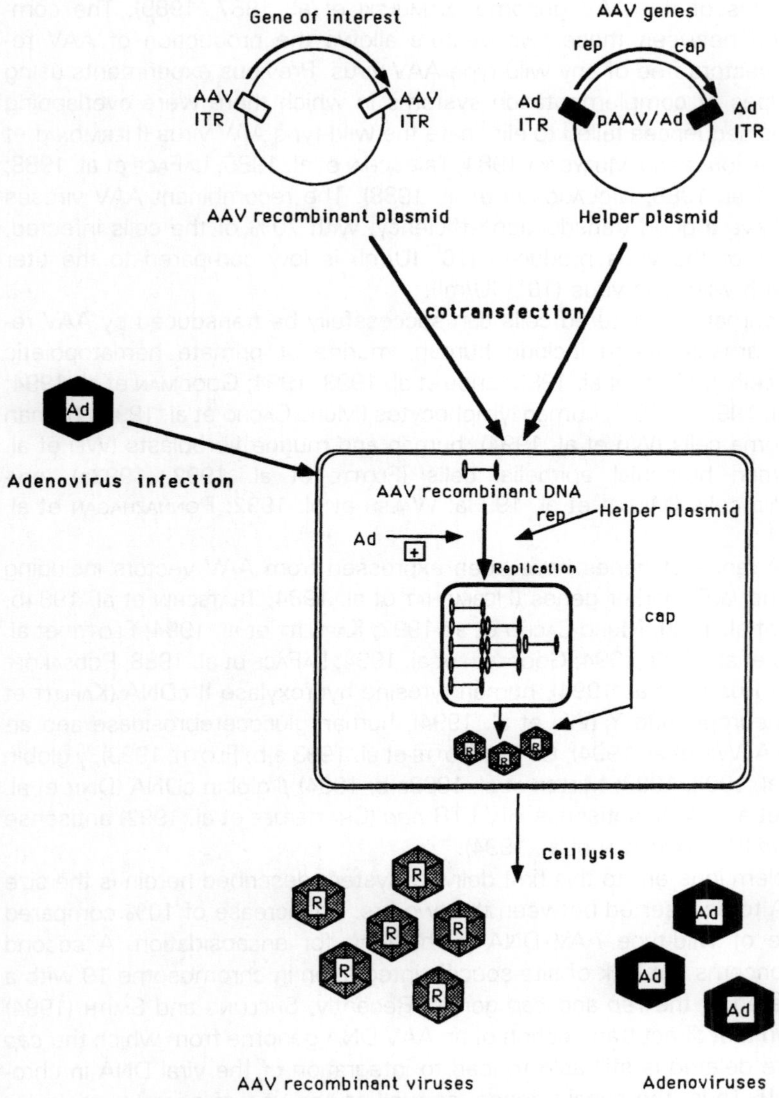

Fig. 7. Production of adeno-associated virus (AAV) recombinant vectors. Stocks of AAV recombinant vectors carrying a gene of interest are produced by cotransfection of two plasmids and Ad infection. The helper plasmid (pAAV/Ad) expresses the *cap* and *rep* genes necessary for rescue, replication and encapsidation of the AAV recombinant DNA carried by the other plasmid. Deprived of its inverted terminal repeat, the AAV DNA carried by the helper plasmid cannot be rescued, replicated and encapsidated in a virus form

and *rep* genes of the AAV genome (SAMULSKI et al. 1987, 1989). The complementation between these two vectors allows the production of AAV recombinant vectors free of any wild-type AAV virus. Previous experiments using different types of complementation systems in which there were overlapping homologous sequences failed to eliminate the wild-type AAV virus (HERMONAT et al. 1984; HERMONAT and MUZYCZKA 1984; TRATSCHIN et al. 1985; LAFACE et al. 1988; LEBKOWSKI et al. 1988; MCLAUGHLIN et al. 1988). The recombinant AAV viruses appear to have a good transduction efficiency, with 70% of the cells infected, but the titer of the virus produced (10^6 IU/ml) is low compared to the titer obtained with wild-type virus (10^{11} IU/ml).

Many primary or cultured cells can successfully be transduced by AAV recombinant viruses: these include human, murine or primate hematopoietic progenitor cells (LAFACE et al. 1988; ZHOU et al. 1993, 1994; GOODMAN et al. 1994; MILLER et al. 1993b, 1994), human lymphocytes (MURO-CACHO et al. 1992), human neuroblastoma cells (WU et al. 1994), human and murine fibroblasts (WEI et al. 1994), human bronchial epithelial cells (FLOTTE et al. 1992, 1993a), erythroleukemia cells (MILLER et al. 1993a; WALSH et al. 1992; PONNAZHAGAN et al. 1994).

A wide range of genes have been expressed from AAV vectors including *neo, cat,* and *lac*Z marker genes (HERMONAT et al. 1984; TRATSCHIN et al. 1984b; LEBKOWSKI et al. 1988; MURO-CACHO et al. 1992; KAPPLITT et al. 1994; FLOTTE et al. 1992; ZHOU et al. 1993, 1994; GOODMAN et al. 1994; LAFACE et al. 1988; PODSAKOFF et al. 1994; RUSSEL et al. 1994), human tyrosine hydroxylase II cDNA (KAPLITT et al. 1994), neuropeptide Y (WU et al. 1994), human glucocerebrosidase and arylsulfatase A (WEI et al. 1994), CFTR (FLOTTE et al. 1993 a,b; FLOTTE 1993), γ globin (WALSH et al. 1992, 1993; MILLER et al. 1993a,b, 1994) β globin cDNA (DIXIT et al. 1991; OHI et al. 1989), antisense HIV-LTR *neo* (CHATTERJEE et al. 1992) antisense α-globin *neo* (PONNAZHAGAN et al. 1994).

A problem inherent to this first delivery system described herein is the size of the DNA to be inserted between the two itrs. An increase of 10% compared to the size of wild-type AAV DNA is the limit for encapsidation. A second problem concerns the lack of site-specific integration in chromosome 19 with a vector deleted of the *rep* and *cap* genes. Recently, SHELLING and SMITH (1994) have shown that direct transfection of an AAV DNA genome from which the *cap* genes were deleted is still able to lead to integration of the viral DNA in chromosome 19. Thus, the capsid genes, as well as the use of a viral particle for transduction, do not appear to be required for site-specific integration of a recombinant vector. Since the rep gene seems necessary for site specific integration, other kinds of vectors and delivery system have to be considered. Recently, an AAV vector carrying the human interleukin-2 (IL2) gene was transfected via liposomes into primary T lymphocytes and primary and cultured tumor cells (breast, ovarian and lung) (PHILIP et al. 1994). Creation of cell lines expressing the complementing *cap* and *rep* genes has been studied. A cell line was made (VINCENT et al. 1990) but either wild-type virus was produced or a very low titer of recombinant virus was made. A cell line efficiently expressing the *rep* genes (YANG et al. 1994) has been made, but some problems due to the toxicity of these genes in cell culture must be considered.

Another important aspect of a vector is the ability to infect and integrate in nondividing cells (e.g., stem cells or neurons). The inability to integrate in non-dividing cells has been a major drawback for retrovirus vectors. Although this point is not absolutely confirmed for AAV, data which have been reported do indicate that AAV can integrate in nondividing cells. One experiment which has been reported involved infection of cells in culture in which cell division had been blocked by a metabolic inhibitor (PODSAKOFF et al. 1994). Such cells could be transformed by AAV vectors; however, the caveat in this type of experiment is that the block to cell inhibition had to be removed for the assessment of the selectable marker used as a reporter gene. In a recent report this type of potential criticism was overcome by using a reporter gene for which expression could be assayed without allowing cell division to occur (RUSSELL et al. 1994). From the results of this experiment, it was concluded that AAV could integrate in cells which were not dividing, but at much lower frequency, approximately 0.5% of that observed in dividing cells. This report also confirmed earlier reports that the entering AAV genome was stable prior to integration for at least 4–6 days. Thus, AAV does appear to have a significant advantage as a vector with respect to being able to transform nondividing stem cells and neurons.

A recent in vivo experiment has shown the possibility of infecting adult rat brain cells with a recombinant AAV virus. The expression of the lac Z gene and the human tyrosine hydrolase was observed for at least 2–3 months (KAPLITT et al. 1994). Another in vivo experiment was also reported (FLOTTE et al. 1993b) with the expression of the CFTR gene in a single lobe of the New Zealand white rabbit lung.

In summary, the AAV recombinant vectors used until now (Fig. 7) have shown a broad host range of infectivity including nondividing cells and a stable expression of the foreign genes in an integrated state. If the attractive feature of preferential integration in chromosome 19 q seems to be lost with a first generation of recombinant vectors, it is still possible that the vectors are integrated in a limited number of sites.

The remainder of the chapters in the volume describe the use of AAV as a vector in terms of various approaches to vector construction, the types of constructs which have been made, and the cells and animal models which have been transformed by the vectors. There are many questions which remain to be answered. These include: (1) Is site specific integration desirable? Potentially, knowledge of the site of integration is a powerful asset. Presumably this is the site used for wild-type integration, which has not been demonstrated to have negative or harmful consequences in the intact host. Even in cell culture, the effects are minimal; however, they maybe detected and in several instances imply an increased sensitivity to genotoxic stimuli. In every latently infected cell line which has been characterized to date, at least one copy of the normal cellular sequence remains intact. When latently infected cells are super-infected with marked AAV, integration is found in alternative loci in the genome of those cells which can be propagated (Samulski, personal communication). This intriguing result suggests the possibility that cell survival requires at least one intact copy of the sequence of which aavS1 is a component. (2) Is the specific site of integration (aavS1) optimal for expression of all genes which might be transferred? (3) To achieve site specific integration Rep is required. If the *rep* gene is present, can it be regulated in a way that does not lead to overexpression that might be toxic to the cell? Under normal conditions this does not seem to be a major problem, although as stated above there may be subtle changes to the phenotype of the cell. Another difficulty is that the capacity of AAV as a virion vector is limited to about 4 kb by packaging considerations. If *rep* is present, then the carrying capacity is reduced by half.

A last concern of some is whether superinfection of a cell by a helper adeno- or herpesvirus might rescue the vector genome from the integrated state. It should be remembered that such an infection will be lethal for the.cell in almost every case and that rescue will not be effected in the absence of a *rep* which can be activated.

In summary, AAV as a virion vector may be satisfactory for gene transfer in some cases. In others it may be desirable to construct synthetic vectors which might be hybrids with adenovirus or constructs carried in liposomes or by other inert vectors. Such new approaches to vector construction might be able to overcome concerns about unregulated *rep* expression or the limited carrying capacity of the AAV genome.

References

Atchinson RW, Casto BC, Hammond WM (1965) Adenovirus-associated defective virus particles. Science 149: 754–756

Bantel-Schaal U (1991) Infection with adeno-associated parvovirus leads to increased sensitivity of mammalian cells to stress. Virology 182: 260–268

Bantel-Schaal U, Stöhr M (1992) Influence of adeno-associated virus on adherence and growth properties of normal cells. J Virol 66: 773–779

Beaton A, Palumbo P, Berns KI (1989) Expression from the adeno-associated virus p5 and p19 promoters is negatively regulated in trans by the rep protein. J Virol 63: 4450–4454

Becerra SP, Koczot F, Fabisch P, Rose JA (1988) Synthesis of adeno-associated virus structural proteins requires both alternative mRNA splicing and alternative initiations form a single transcript. J Virol 62: 2745–2754

Berns KI (1990a) Parvoviridae and their replication. In: Fields BN, Knipe DM (eds) Virology, vol 2, 2nd edn. Raven, New York, pp 1743–1763

Berns KI (1990b) Parvovirus replication. Mircobiol Rev 54: 316–329

Berns KI, Linden RM (1995) The cryptic life cycle of adeno-associated virus. Bioessays 17: 237–245

Berns KI, Pinkerton TC, Thomas GF, Hoggan MD (1975) Detection of adeno-associated virus (AAV) specific nucleotide sequences in DNA isolated from latently infected Detroit 6 cells. Virology 68: 556–560

Berns KI, Kotin RM, Labow MA (1988) Regulation of adeno-associated virus DNA replication. Biochem Biophys Acta 951: 425–429

Bohenzky RA, Le Febvre RB, Berns KI (1988) Sequence and symmetry requirements within the internal palindromic sequences of the adeno-associated virus terminal repeat. Virology 166: 316–327

Buller RM, Janik JE, Sebring ED, Rose JA (1981) Herpes simplex virus type 1 and 2 completely help adeno-associated virus replication. J Virol 40: 241–247

Carter BJ (1992) Adeno-associated virus vectors. Curr Opin Biotechnol 3: 533–539

Cassinotti P, Weitz M, Tratschin JD (1988) Organization of the adeno-associated virus (AAV) capsid gene: mapping of a minor spliced mRNA coding for virus capsid protein 1. Virology 167: 176–184

Chang LS, Shi Y, Shenk T (1989) Adeno-associated virus p5 promoter contains an adenovirus E1A inducible element and a binding site for the major late transcription factor. J Virol 63: 3479–3488

Chatterjee S, Johnson PR, Wong KK (1992) Dual-target inhibition of HIV-1 in vitro by means of an adeno-associated virus antisense vector. Science 258: 1485–1488

Chejanovsky N, Carter BJ (1989) Mutagenesis of an AUG codon in the adeno-associated virus rep gene: effects on viral DNA replication. Virology 173: 120–128

Cheung AK, Hoggan MD, Hauswirth WW, Berns KI (1980) Integration of the adeno-associated virus genome into cellular DNA in latently infected human Detroit 6 cells. J Virol 33: 739–748

Chiorini JA, Weitzman MD, Owens RA, Urcelay E, Safer B, Kotin RM (1994) Biologically active Rep proteins of adeno-associated virus type 2 produced as fusion proteins in Escherichia coli. J Virol 68: 797–804.

Das HK, Jackson CL, Miller DA, Leff T, Breslow JL (1987) The human apolipoprotein C-II gene sequence contains a novel chromosome 19– specific minisatellite in its third intron. J Biol Chem 262: 4787–4793

Dixit M, Webb MS, Smart WC, Ohi S (1991) Construction and expression of a recombinant adeno-associated virus that harbors a human beta-globin-encoding cDNA. Gene 104: 253–257

Flotte TR (1993) Prospects for virus-based gene therapy for cystic fibrosis. J Bioenerg Biomembr 25: 37–42

Flotte TR, Solow R, Owens RA, Afione S, Zeitlin PL, Carter BJ (1992) Gene expression from adeno-associated virus vectors in airway epithelial cells. Am J Respir Cell Mol Biol 7: 349–356

Flotte TR, Afione SA, Solow R, Drumm ML, Markakis D, Guggino WB, Zeitlin PL, Carter BJ (1993a) Expression of the cystic fibrosis transmembrane conductance regulator from a novel adeno-associated virus promoter. J Biol Chem 268: 3781–3790

Flotte TR, Afione SA, Conrad C, McGrath SA, Solow R, Oka H, Zeitlin PL, Guggino WB, Carter BJ (1993b) Stable in vivo expression of the cystic fibrosis transmembrane conductance regulator with an adeno-associated virus vector. Proc Natl Acad Sci USA 90: 10613–10617

Giraud C, Winocour E, Berns KI (1994) Site-specific integration by adeno-associated virus is directed by a cellular DNA sequence. Proc Natl Acad Sci USA 91: 10039–10043

Giraud C, Winocour E, Berns KI (1995) Recombinant junctions formed by site-specific integration of adeno-associated virus into an episome. J Virol 69: 6917–6924

Goodman S, Xiao X, Donahue RE, Moulton A, Miller J, Walsh C, Young NS, Samulski RJ, Nienhuis AW (1994) Recombinant adeno-associated virus-mediated gene transfer into hematopoietic progenitor cells. Blood 84: 1492–1500

Handa H, Shiroki K, Shimojo H (1977) Establishment and characterization of KB cells lines latently infected with adeno-associated virus type 1. Virology 82: 84–92

Hauswirth WW, Berns KI (1979) Adeno-associated virus DNA replication: non unit-length molecules. Virology 93: 57–68

Hermonat PL, Muzyczka N (1984) Use of adeno-associated virus as a mammalian DNA cloning vector: transduction of neomycin resistance into mammalian tissue culture cells. Proc Natl Acad Sci USA 81: 6466–6470

Hermonat PL, Labow MA, Wright R, Berns KI, Muzyczka N (1984) Genetics of adeno-associated virus: isolation and preliminary characterization of adeno-associated virus type 2 mutants. J Virol 51: 329–339

Hoggan MD, Blacklow NR, Rowe WP (1966) Studies of small DNA viruses found in various adenovirus preparations: physical, biological, and immunological characteristics. Proc Natl Acad Sci USA 55: 1457–1471

Hoggan MD, Thomas GF, Thomas FB, Johnson FB (1972) Continuous "carriage" of adenovirus associated genome in cell cultures in the absence of helper adenoviruses. Proceedings of the 4th Lepetit Colloquium. Cocoyac, Mexico. North Holland Publishing Co., Amsterdam

Hong G, Ward P, Berns KI (1994) Intermediates of adeno-associated virus DNA replication in vitro. J Virol 68: 2011–2015

Im DS, Muzyczka N (1989) Factors that bind to adeno-associated virus terminal repeats. J Virol 63: 3095–3104

Im DS, Muzyczka N (1990) The AAV origin binding protein Rep68 is an ATP-dependent site-specific endonuclease with DNA helicase activity. Cell 61: 447–457
Im DS, Muzyczka N (1992) Partial purification of adeno-associated virus Rep78, Rep52, and Rep40 and their biochemical characterization. J Virol 66: 1119–1128
Janik JE, Huston MM, Cho K, Rose JA (1989) Efficient synthesis of adeno-associated virus structural proteins requires both adenovirus DNA binding protein and VA I RNA. Virology 168: 320–329
Jay FT, de la Maza LM, Carter BJ (1979) Parvovirus RNA transcripts containing sequences not present in mature mRNA: a method for isolation of putative mRNA precursor sequences. Proc Natl Acad Sci USA 76: 625–629
Kaplitt MG, Leone P, Samulski RJ, Xiao X, Pfaff DW, O'Malley KL, During MJ (1994) Long-term gene expression and phenotypic correction using adeno-associated virus vectors in the mammalian brain. Nature Genet 8: 148–154
Klein-Bauernschmitt P, zur Hausen H, Schlehofer JR (1992) Induction of differentiation-associated changes in established human cells by infection with adeno-associated virus type 2. J Virol 66: 4191–4200
Kotin RM (1994) Prospects for the use of adeno-associated virus as a vector for human gene therapy. Hum Gene Ther 5: 793–801
Kotin RM, Berns KI (1989) Organization of adeno-associated virus DNA in latently infected Detroit 6 cells. Virology 170: 460–467
Kotin RM, Siniscalco M, Samulski RJ, Zhu XD, Hunter L, Laughlin CA, McLaughlin S, Muzyczka N, Rocchi M, Berns KI (1990) Site-specific integration by adeno-associated virus. Proc Natl Acad Sci USA 87: 2211–2215
Kotin RM, Menninger JC, Ward DC, Berns KI (1991) Mapping and direct visualization of a region-specific viral DNA integration site on chromosome 19q13-qter. Genomics 10: 831–834
Kotin RM, Linden RM, Berns KI (1992) Characterization of a preferred site on human chromosome 19q for integration of adeno-associated virus DNA by non-homologous recombination. EMBO J 11: 5071–5078
Labow MA, Berns KI (1988) The adeno-associated virus rep gene inhibits replication of an adeno-associated virus/simian virus 40 hybrid genome in cos-7 cells. J Virol 62: 1705–1712
Labow MA, Hermonat PL, Berns KI (1986) Positive and negative autoregulation of the adeno-associated virus type 2 genome. J Virol 60: 251–258
LaFace D, Hermonat P, Wakeland E, Peck A (1988) Gene transfer into hematopoietic progenitor cells mediated by an adeno-associated virus vector. Virology 162: 483–486
Laughlin CA, Tratschin JD, Coon H, Carter BJ (1983) Cloning of infectious adeno-associated virus genomes in bacterial plasmids. Gene 23: 65–73
Laughlin CA, Cardellichio CB, Coon HC (1986) Latent infection of KB cells with adeno-associated virus type 2. J Virol 60: 515–524
Lebkowski JS, McNally MM, Okarma TB, Lerch LB (1988) Adeno-associated virus: a vector system for efficient introduction and integration of DNA into a variety of mammalian cell types. Mol Cell Biol 8: 3988–3996
Leonard CJ, Berns KI (1994a) Cloning, expression, and partial purification of Rep78: an adeno-associated virus replication protein. Virology 200: 566–573
Leonard CJ, Berns KI (1994b) Adeno-associated virus type 2: a latent life cycle. Prog Nucleic Acid Res Mol Biol 48: 29–52
Margolskee RF (1992) Epstein-Barr virus based expression vectors. In: Muzyczka N (ed) Viral expression vectors. Springer, Berlin Heidelberg New York, pp 67–95 (Current topics in microbiology and immunology vol 158)
McCarty DM, Pereira DJ, Zolotukhin I, Zhou X, Ryan JH, Muzyczka N (1994a) Identification of linear DNA sequences that specifically bind the adeno-associated virus rep protein. J Virol 68: 4988–4997
McCarty DM, Ryan JH, Zolotukhin S, Zhou X, Muzyczka N (1994b) Interaction of the adeno-associated virus rep protein with a sequence within the A palindrome of the viral terminal repeat. J Virol 68: 4998–5006
McLaughlin SK, Collis P, Hermonat PL, Muzyczka N (1988) Adeno-associated virus general transduction vectors: analysis of proviral structures. J Virol 62: 1963–1973
Mendelson E, Trempe JP, Carter BJ (1986) Identification of the transacting Rep proteins of adeno-associated virus by antibodies to a synthetic oligopeptide. J Virol 60: 823–832
Miller JL, Walsh CE, Ney PA, Samulski RJ, Nienhuis AW (1993a) Single-copy transduction and expression of human γ-globin in K562 erythroleukemia cells using recombinant adeno-associated

virus vectors: the effect of mutations in NF-E2 and GATA-1 binding motifs within the hypersensitivity site 2 enhancer. Blood 82: 1900–1906

Miller JL, Walsh CE, Samulski RJ, Young NS, Nienhuis AW (1993b) Transfer and expression of human-γ-globin gene in purified hematopoeitic progenitor cells from Rhesus bone marrow using a recombinant adeno-associated viral (rAAV) vector (abstract). 5th parvovirus workshop, Cristal River, FL, USA

Miller JL, Donahue RE, Sellers SE, Samulski RJ, Young NS, Nienhuis AW (1994) Recombinant adeno-associated virus (rAAV)-mediated expression of human g-globin gene in human progenitor-derived erythroid cells. Proc Natl Acad Sci USA 91: 10183–10187

Mishra L, Rose JA (1990) Adeno-associated virus DNA replication is induced by genes that are essential for HSV-1 DNA synthesis. Virology 179: 632–639

Muralidhar S, Becerra SP, Rose JA (1994) Site-directed mutagenesis of adeno-associated virus type 2 structural protein initiation codons: effects on regulation of synthesis and biological activity. J Virol 68: 170–176

Muro-Cacho CA, Samulski RJ, Kaplan D (1992) Gene transfer in human lymphocytes using a vector based on adeno-associated virus. J Immunother 11: 231–237

Muzyczka N (1992) Use of adeno-associated virus as a general transduction vector for mammalian cells. In: Muzyczka N (ed) Viral expression vectors. Springer, Berlin Heidelberg New York, pp 97–129 (Current topics in microbiology and immunology, vol 158)

Nevins JR (1982) Induction of the synthesis of a 70,000 dalton mammalian heat shock protein by the adenovirus E1A product. Cell 29: 213–220

Ni TH, Zhou X, McCarty DM, Zolotukhin I, Muzyczka N (1994) In vitro replication of adeno-associated virus DNA. J Virol 68: 1128–1138

Ohi S, Dixit M, Tillery MK, Plonk SG (1989) Construction and replication of an adeno-associated virus expression vector that contains human beta-globin cDNA. Gene 89: 279–282

Philip R, Brunette E, Kilinski L, Murugesh D, NcNally MA, Ucar K, Rosenblatt J, Okarma TB, Lebkowski JS (1994) Efficient and sustained gene expression in primary T lympocytes and primary and cultured tumor cells mediated by adeno-associated virus plasmid DNA complexed to cationic liposomes. Mol Cell Biol 14: 2411–2418

Podsakoff G, Wong K Jr, Chatterjee S (1994) Efficient gene transfer into nondividing cells by adeno-associated virus-based vectors. J Virol 68: 5656–5666

Ponnazhagan S, Nallari ML, Srivastava A (1994) Suppression of human α-globin gene expression mediated by the recombinant adeno-associated virus 2-based antisense vectors. J Exp Med 179: 733–738

Ponticelli SA, Smith GR (1992) Chromosomal context dependence of a eukariotic recombinational hot spot. Proc Natl Acad Sci USA 89: 227–231

Richardson WD, Westphal H (1981) A cascade of adenovirus early functions is required for expression of adeno-associated virus. Cell 27: 133–141

Richardson WD, Westphal HC (1984) Requirement for either early region 1a or early region 1b adenovirus gene products in the helper effect for adeno-associated virus. J Virol 51: 404–410

Richardson WD, Carter BJ, Westphal H (1980) Vero cells injected with adenovirus type 2 mRNA produce authentic viral polypeptide patterns: early mRNA promoted growth of adeno-associated virus. Proc Natl Acad Sci USA 77: 931–935

Russell DW, Miller AD, Alexander IE (1994) Adeno-associated virus vectors preferentially transduce cells in S phase. Proc Natl Acad Sci USA 91: 8915–8919

Samulski RJ, Berns KI, Tan M, Muzyczka N (1982) Cloning of adeno-associated virus into pBR322: rescue of intact virus from the recombinant plasmid in human cells. Proc Natl Acad Sci USA 79: 2077–2081

Samulski RJ (1993) Adeno-associated virus: integration at a specific chromosomal locus. Curr Opin Genet Dev 3: 74–80

Samulski RJ, Shenk T (1988) Adenovirus E1B 55-Mr polypeptide facilitates timely cytoplasmic accumulation of adeno-associated virus mRNAs. J Virol 62: 206–210

Samulski RJ, Srivastava A, Berns KI, Muzyczka N (1983) Rescue of adeno-associated virus from recombinant plasmids: gene correction within the terminal repeats of AAV. Cell 33: 135–143

Samulski RJ, Chang L-S, Shenk T (1987) A recombinant plasmid from which an infectious adeno-associated virus genome can be excised in vitro and its use to study viral replication. J Virol 61: 3096–3101

Samulski RJ, Chang L-S, Shenk T (1989) Helper-free stocks of recombinant adeno-associated viruses: normal integration does not require viral gene expression. J Virol 63: 3822–3828

Samulski RJ, Zhu X, Xiao X, Brook JD, Housman DE, Epstein N, Hunter LA (1991) Targeted integration of adeno-associated virus (AAV) into human chromosome 19. EMBO J 10: 3941–3950
Schlehofer JR, Ehrbar M, zur Hausen H (1986) Vaccinia virus, herpes simplex virus, and carcinogens induce DNA amplification in a human cell line and support replication of a helpervirus dependent parvovirus. Virology 152: 110–117
Schuchert P, Langsford M, Käslin E, Kohli J (1991) A specific DNA sequence is required for high frequency of recombination in the ade 6 gene of fission yeast. EMBO J 10: 2157–2163
Senapathy P, Tratschin JD, Carter BJ (1984) Replication of adeno-associated virus DNA. Complementation of naturally occurring rep-mutants by a wild-type genome or an ori- mutant and correction of terminal palindrome deletions. J Mol Biol 179: 1–20
Shelling AN, Smith MG (1994) Targeted integration of transfected and infected adeno-associated virus vectors containing the neomycin resistant gene. Gene Ther 1: 165–169
Siegl G, Bates RC, Berns KI, Carter BJ, Kelly DC, Kurstak E, Tattersall P (1985) Characteristics and taxonomy of parvoviridae. Intervirology 23: 61–73
Snyder RO, Im D-S, Ni T, Xiao X, Samulski RJ, Muzyczka N (1993) Features of the adeno-associated virus origin involved in substrate recognition by the viral rep protein. J Virol 67: 6096–6104
Srivastava A (1987) Replication of the adeno-associated virus DNA termini in vitro. Intervirology 27: 138–147
Srivastava A, Lusby EW, Berns KI (1983) Nucleotide sequence and organization of the adeno-associated virus 2 genome. J Virol 45: 555–564
Stillman BW, Lewis JB, Chow LT, Matthews MB, Smart JE (1981) Identification of the gene for the adenovirus terminal protein precursor. Cell 23: 497–508
Stillman BW, Tamonoi F, Matthews MB (1982) Purification of an adenovirus-coded DNA polymerase that is required for initiation of DNA replication. Cell 31: 613–623
Straus SE, Sebring E, Rose JA (1976) Concatemers of alterning plus and minus strands are intermediates in adeno-associated virus DNA synthesis. Proc Natl Acad Sci USA 73: 742–746
Tratschin JD, Miller IL, Carter BJ (1984a) Genetic analysis of adeno-associated virus: properties of deletion mutants constructed in vitro and evidence for an adeno-associated virus replication function. J Virol 51: 611–619
Tratschin JD, West MH, Sandbank T, Carter BJ (1984b) A human parvovirus, adeno-associated virus, as a eucaryotic vector: transient expression and encapsidation of the procaryotic gene for chloramphenicol acetyltransferase. Mol Cell Biol 4: 2072–2081
Tratschin JD, Miller IL, Smith MG, Carter BJ (1985) Adeno-associated virus vector for high-frequency integration, expression, and rescue of genes in mammalian cells. Mol Cell Biol 5: 3251–3260
Tratschin JD, Tal J, Carter BJ (1986) Negative and positive regulation in trans of gene expression from adeno-associated virus vectors in mammalian cells by a viral rep gene product. Mol Cell Biol 6: 2884–2894
Trempe JP, Carter BJ (1988) Alternate mRNA splicing is required for synthesis of adeno-associated virus VP1 capsid protein. J Virol 62: 3356–3363
Urcelay E, Ward P, Wiener SM, Safer B, Kotin RM (1995) Asymmetric replication in vitro from a human sequence element is dependent on adeno-associated virus rep protein. J Virol 69: 2038–2046
Vincent KA, Moore GK, Haigwood NL (1990) Replication and packaging of HIV envelope genes in a novel adeno-associated virus vector system. Vaccines 90: 353–359
Walsh CE, Liu JM, Xiao X, Young NS, Nienhuis AW, Samulski RJ (1992) Regulated high level expression of a human gamma-globin gene introduced into erythroid cells by an adeno-associated virus vector. Proc Natl Acad Sci USA 89: 7257–7261
Walsh CE, Liu JM, Miller JL, Nienhuis AW, Samulski RJ (1993) Gene therapy for human hemoglobinopathies. Proc Soc Exp Biol Med 204: 289–300
Walz C, Schlehofer JR (1992) Modification of some biological properties of HeLa cells containing adeno-associated virus DNA integrated into chromosome 17. J Virol 66: 2990–3002
Walz C, Schlehofer JR, Flentje M, Rudat V, zur Hausen H (1992) Adeno-associated virus sensitizes HeLa cell tumors to gamma rays. J Virol 66: 5651–5657
Ward P, Urcelay E, Kotin R, Safer B, Berns KI (1994) Adeno-associated virus DNA replication in vitro: activation by a maltose binding protein/Rep 68 fusion protein. J Virol 68: 6029–6037
Wei J-F, Wei F-S, Samulski RJ, Barranger JA (1994) Expression of the human glucocerebrosidase and arylsulfatase A genes in murine and patient primary fibroblasts transduced by an adeno-associated virus vector. Gene Ther 1: 261–268
Weindler FW, Heilbronn R (1991) A subset of herpes simplex virus replication genes provides helper functions for productive adeno-associated virus replication. J Virol 65: 2476–2483

Weitzman MD, Kyostio SR, Kotin RM, Owens RA (1994) Adeno-associated virus (AAV) Rep proteins mediate complex formation between AAV DNA and its integration site in human DNA. Proc Natl Acad Sci USA 91: 5808–5812

West MH, Trempe JP, Tratschin JD, Carter BJ (1987) Gene expression in adeno-associated virus vectors: the effects of chimeric mRNA structure, helper virus, and adenovirus VA1 RNA. Virology 160: 38–47

Winocour E, Callaham MF, Huberman E (1988) Perturbation of the cell cycle by adeno-associated virus. Virology 167: 393–399

Winocour E, Puzis L, Etkin S, Koch T, Danovitch B, Mendelson E, Shaulian E, Karby S, Lavi S (1992) Modulation of the cellular phenotype by integrated adeno-associated virus. Virology 190: 316–329

Wu BJ, Hurst HC, Jones NC, Morimoto RJ (1986) The E1A 13S product of adenovirus 5 activates transcription of the cellular human HSP70 gene. Mol Cell Biol 6: 2994–2999

Wu P, Ziska D, Bonell MA, Grouzmann E, Millard WJ, Meyer EM (1994) Differential neuropeptide Y gene expression in post-mitotic versus dividing neuroblastoma cells driven by an adeno-associated virus vector. Mol Brain Res 24: 27–33

Xiao X, deVlaminck W, Monahan J (1993) Adeno-associated virus (AAV) vectors for gene transfer. Adv Drug Deliv Rev 12: 201–215

Yakobson B, Koch T, Winocour E (1987) Replication of adeno-associated virus in synchronized cells without the addition of a helper virus. J Virol 61: 972–981

Yakobson B, Hrynko TA, Peak MJ, Winocour E (1989) Replication of adeno-associated virus in cells irradiated with UV light at 254 nm. J Virol 63: 1023–1030

Yalkinoglu AO, Heilbronn R, Burkle A, Schlehofer JR, zur Hausen H (1988) DNA amplification of adeno-associated virus as a response to cellular genotoxic stress. Cancer Res 48: 3123–3129

Yang Q, Chen F, Trempe JP (1994) Characterization of cell lines that inducibly express the adeno-associated virus rep proteins. J Virol 68: 4847–4856

Zhou SZ, Broxmeyer HE, Cooper S, Harrington MA, Srivastava A (1993) Adeno-associated virus 2-mediated gene transfer in murine hematopoieteic progenitor cells. Exp Hematol 21: 928–933

Zhou SZ, Cooper S, Kang LY, Ruggieri L, Heimfeld S, Srivastava A, Broxmeyer HE (1994) Adeno-associated virus 2-mediated high efficiency gene transfer into immature and mature subsets of hematopoietic progenitor cells in human umbilical cord blood. J Exp Med 179: 1867–1875

Zhu X (1993) Characterization of adeno-associated virus proviral structure in latently infected human cells. Doctor of philosophy. Pittsburgh, 1993

The Roles of AAV Rep Proteins in Gene Expression and Targeted Integration

J.A. Chiorini, S.M. Wiener, L. Yang, R.H. Smith, B. Safer, N.P. Kilcoin, Y. Liu, E. Urcelay, and R.M. Kotin

1 Introduction

Adeno-associated virus 2 (AAV), a nonpathogenic human virus, can either integrate into host chromatin and remain latent or replicate following infection. The outcome depends on the cellular conditions. Under conditions permissive for AAV DNA replication (e.g., during adenovirus coinfection), AAV gene expression is induced. The single-stranded AAV genome of 4680 nucleotides is organized into two open reading frames (ORFs) that encode structural capsid proteins (Cap) and nonstructural proteins (Rep) (Fig. 1). Two promoters at AAV map units 5 and 19, p5 and p19, direct expression of the *rep* gene. The *cap* gene is regulated by the p40 promoter. A common intron results in the production of four Rep proteins: p5 initiated Rep78, Rep68 and p19 initiated Rep52, Rep40. The inverted terminal repeats (ITRs) function as viral origins of replication required for encapsidation of the of AAV DNA. The production of AAV Rep proteins enables viral DNA to replicate, resulting in a geometric increase in the number of viral genomes. The AAV p5 initiated regulatory proteins Rep78 and Rep68 interact with the viral promoters to establish a feedback loop. These Rep proteins are involved directly in viral DNA replication as well.

In contrast, latent infection results from AAV infection of cells nonpermissive for viral replication. Viral DNA stably integrates into the cellular genome and, in latent infections established in tissue culture, the provirus DNA integrates into a defined region of human chromosome 19 (Kotin et al. 1990,

Molecular Hematology Branch MSC 1654, National Heart, Lung, and Blood Institute, National Institutes of Health, 10 Center Drive, Bethesda, MD 20892-1654, USA

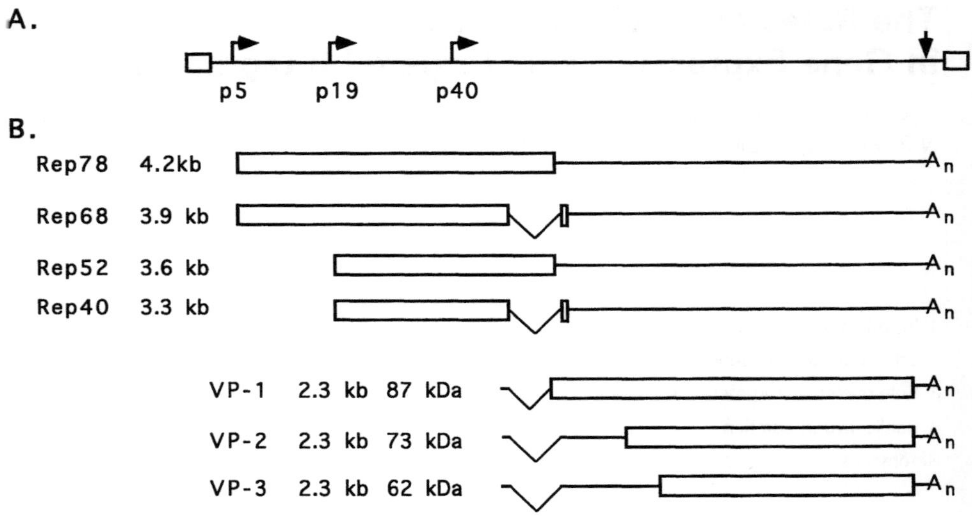

Fig. 1A,B. The physical and genetic map of the adeno-associated virus (AAV) genome. The genome is 4680 nucleotides (nts) which is divided into 100 map units (mu) (From KOTIN 1994). **A** The viral genome. The *open boxes* on both ends indicate the positions of the inverted terminal repeats (itrs). The positions of the three viral promoters at map positions 5, 19 and 40 are indicated by the *right angled arrows*. The site of the common polyadenylation signal at map position 96 is indicated by the *vertical arrow*. **B** Transcripts derived from each of the promoters. The open reading frames are represented by the *open rectangles*, the untranslated regions are indicated by a *solid line* and the introns are shown as a *carat*. The rep transcripts produced from p5 and p19 are present in spliced and unspliced forms. The capsid protein VP-1 utilizes an alternate splice acceptor at nt 2201. VP-2 and VP-3 are derived from the same transcription unit. VP-2 utilizes an ACG triplet as a initiation codon upstream from the start of VP-3

1991, 1992; SAMULSKI et al. 1991). If the cell becomes permissive for viral replication (e.g., when adenovirus superinfects the cell), the proviral *rep* and *cap* genes are expressed and viral replication proceeds.

2 Effects of Rep on Gene Regulation

Regulation of AAV gene expression is complex and involves positive and negative regulation of viral transcription. The two larger nonstructural proteins of AAV, Rep68 and Rep78, are required for this regulation. These proteins have been shown to bind to ITRs (ASHKTORAB and SRIVASTAVA 1989; CHIORINI et al. 1994a; IM and MUZYCZKA 1989, 1990, 1992; OWENS and CARTER 1992; OWENS et al. 1991, 1993; SRIVASTAVA et al. 1983) and stimulate replication both in vivo (BERNS et al. 1988; HERMONAT et al. 1984; LABOW and BERNS 1988) and in vitro (CHIORINI et al. 1994a; HONG et al. 1992; NI et al. 1994). Furthermore, Rep68 and Rep78 have been shown to nick the duplex ITR in a site- and strand-specific manner. This cleavage, which occurs at the terminal resolution site (TRS), permits the replication of the hairpin structure in a process referred to as terminal resolution (IM and MUZYCZKA 1990).

In addition to their role in AAV DNA replication, Rep68 and Rep78 regulate the transcription of the AAV p5, p19, and p40 promoters in vivo (BEATON et al. 1989; KYOSTIO et al. 1994; LABOW et al. 1986; MCCARTY et al. 1994b; TREMPE and CARTER 1988). Rep68 and Rep78 proteins can also inhibit the transcription of heterologous promoters and cellular transformation by papillomavirus or by adenovirus Ela plus an activated *ras* oncogene (HERMONAT 1989, 1991, 1992; KHLEIF et al. 1991; LABOW and BERNS 1988; LABOW et al. 1987).

These cellular effects associated with Rep68 and 78 may result from direct interactions of Rep with cellular DNA. Although the secondary structure of the ITR was considered essential for Rep68 binding (ASHKTORAB and SRIVASTAVA 1989; IM and MUZYCZKA 1990; SNYDER et al. 1993), recent work demonstrated that binding does not require a stem-loop structure and that Rep binds to a linear truncated form of the ITR with affinity similar to that of the wild-type hairpin ITR (CHIORINI et al. 1994a,b). Furthermore, Rep binds to the AAV p5 promoter and to fragments of the cloning vector pBR322 and human chromosome 19 (MCCARTY et al. 1994a; WEITZMAN et al. 1994), two elements which are incapable of assuming a T-shaped ITR structure. However, a common motif in these sequences is a tetranucleotide repeat, GCTC. Gel Shift assays with randomly mutated oligonucleotide fragments established a consensus sequence with a core of two repeated tetranucleotide motifs (unpublished data). Comparison of this consensus sequence with the GenBank database identified several promoters or regulatory regions containing minor variations of this consensus sequence. One such sequence is found within the c-H-*ras* gene promoter. In the presence of Rep, cellular transformation by c-H-*ras* is inhibited (BATCHU et al. 1994). However, other promoters which do not appear to contain a Rep consensus binding element are also inhibited by Rep proteins although not to the same extent as promoters containing the consensus sequence (unpublished). These results indicate that Rep affects gene expression at multiple levels. Specific effects may be attributed to direct Rep: DNA interaction with gene regulatory regions. Thus, these more general Rep activities may directly influence non-AAV gene expression.

In addition to DNA binding, other activities of Rep are not restricted to hairpin ITR substrates. The ITR fragment is a substrate for site-specific endonuclease activity, which is a critical process for viral replication. However, relative to substrates containing a hairpin, endonuclease cleavage of linear substrates was only 1% as efficient, although Rep binds to either substrate with similar affinities (CHIORINI et al. 1994a,b). During a lytic infection, rapid amplification of the viral genome results, necessitating efficient cleavage at the TRS of the ITR. In a nonpermissive cell, cellular linear substrates for Rep cleavage are not undergoing rapid amplification. The inefficient cleavage with the linear substrates may be sufficient for the cellular effects associated with Rep.

Rep effects expression of promoters lacking a Rep binding site. This repression, not attributable to Rep promoter interactions, constitutes a nonspecific effect. Transient expression assays indicate that Rep may also function post-transcriptionally to repress gene expression, as indicated by increased levels of transcripts and a concomitant decrease of reporter gene product activity (unpublished observations; TREMPE and CARTER 1988). Strategies for influencing cellular macromolecular synthesis employed by viruses include affecting message stability or transport (PILDER et al. 1986) or translational initiation (KATZE et al. 1986; KRAUSSLICH et al. 1987).

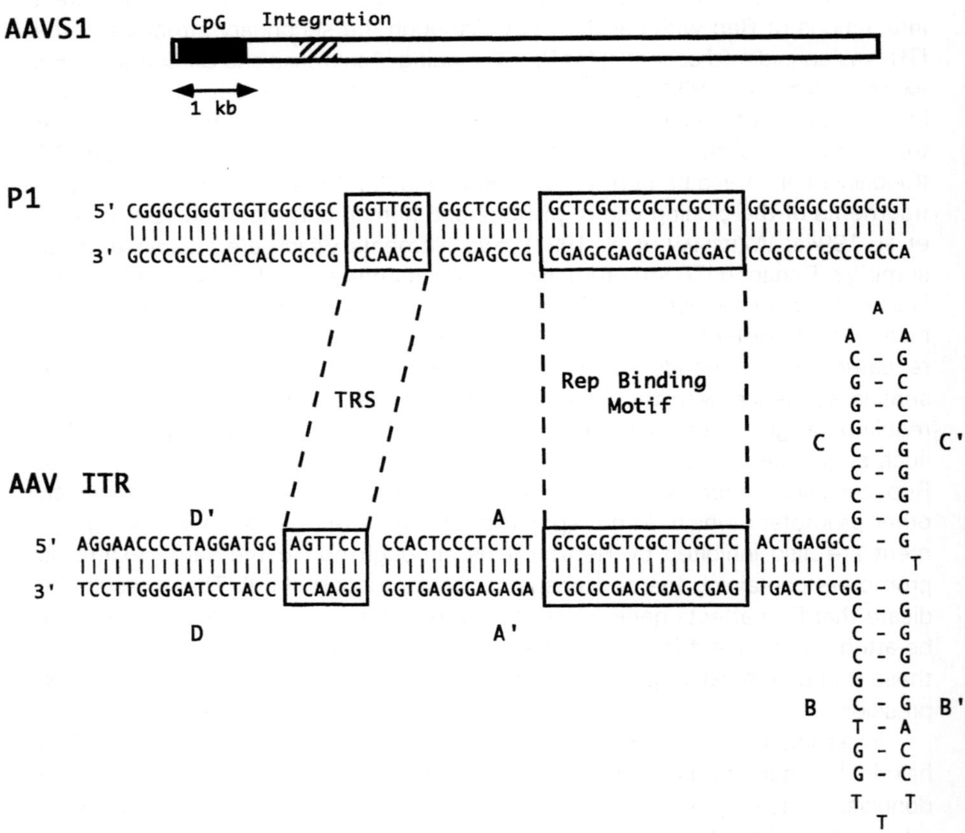

Fig. 2. Sequence alignment of the P1 element derived from the adeno-associated virus chromosome 19 integration locus (AAVS1) with the AAV inverted terminal repeat (*ITR*). AAVS1 is represented schematically with the positions of the CpG island (*filled box*) KOTIN et al. 1992 and sites of proviral integration indicated (*hatched region*) KOTIN et al. 1992, 354–468; SAMULSKI et al. 1991; SHELLING and SMITH 1994. P1 was obtained as a *Sma*I subfragment of AAVS1 (nucleotides KOTIN et al. 1992; see also WEITZMAN et al. 1994). The Rep binding motif of both P1 and AAV ITR are outlined and labeled. The sequences homologous to the terminal resolution site (*TRS*) of AAV are outlined in both the ITR sequence and P1. The endonuclease cleavage site is between the thymidines within the TRS box and is indicated by the *arrow*. (From URCELAY et al. 1995)

3 Role of Rep in Targeted Integration

WEITZMAN et al. (1994) originally described Rep protein recognition of non-AAV DNA. Screening the chromosome 19 integration locus (AAVS1) by electrophoretic mobility shift assay (EMSA) detected a specific interaction between Rep and a 110 bp subfragment of AAVS1 (WEITZMAN et al. 1994). This significant result provided a direct link between Rep and targeted integration of AAV DNA. The sequence of the AAVS1 subfragment, P1, contained a motif homologous to the A region of the AAV ITR (Fig. 2). Methylation inteference experiments indicated that the guanine residue in the GCTC repeat motif is important for Rep binding (WEITZMAN et al. 1994). Furthermore, an element similar to the AAV TRS appeared properly positioned relative to the GCTC repeat motif (IM and MUZYCZKA 1990; URCELAY et al. 1995). Rep cleaved duplex synthetic oligonucleotides corresponding to the P1 sequence predominantly at the predicted site within the TRS homolog (URCELAY et al. 1995). The cleavage efficiency of the synthetic P1 and the truncated ITR was similar (CHIORINI et al. 1994a). In addition, Rep formed a covalent complex with the P1 DNA substrate which is similar to established activity with either hairpinned (IM and MUZYCZKA 1990) or truncated ITR substrates (CHIORINI et al. 1994a).

4 Replicative Integration

Under certain conditions, AAV DNA integrates into a specific region of human chromosome 19 (KOTIN et al. 1990, 1991, 1992; SAMULSKI et al. 1991; SHELLING and SMITH 1994). However, the mechanism of targeted integration remains obscure. Earlier reports, based on Southern blot analysis of latently infected cell lines using AAV DNA-specific probes, detected no extensive sequence identity between AAV and the human genome (CHEUNG et al. 1980; MCLAUGHLIN et al. 1988). Analysis of the cloned integration locus (AAVS1, KOTIN et al. 1991) confirmed that the sequence similarity between the AAVS1 and AAV was limited to short, discrete regions of homology. Sequence comparison of the proviral-cellular junctions revealed a four to five nucleotide identity which may have been donated by either substrate. Therefore, integration occurred by nonhomologous recombination. Sequence-specific DNA binding proteins (Rep) could juxtapose the viral and cellular DNA. This model, diagrammed in Fig. 3, provides the basis for a simple mechanism of targeted integration for AAV.

Results extending the DNA binding and strand-specific endonuclease activities of Rep68 and Rep78 to include linear duplex DNA in general and P1 specifically argue for Rep participation in targeted integration of AAV DNA. In vivo observations correlate higher frequencies of targeted integration with the presence of an intact rep gene (KOTIN et al. 1990) providing further support for

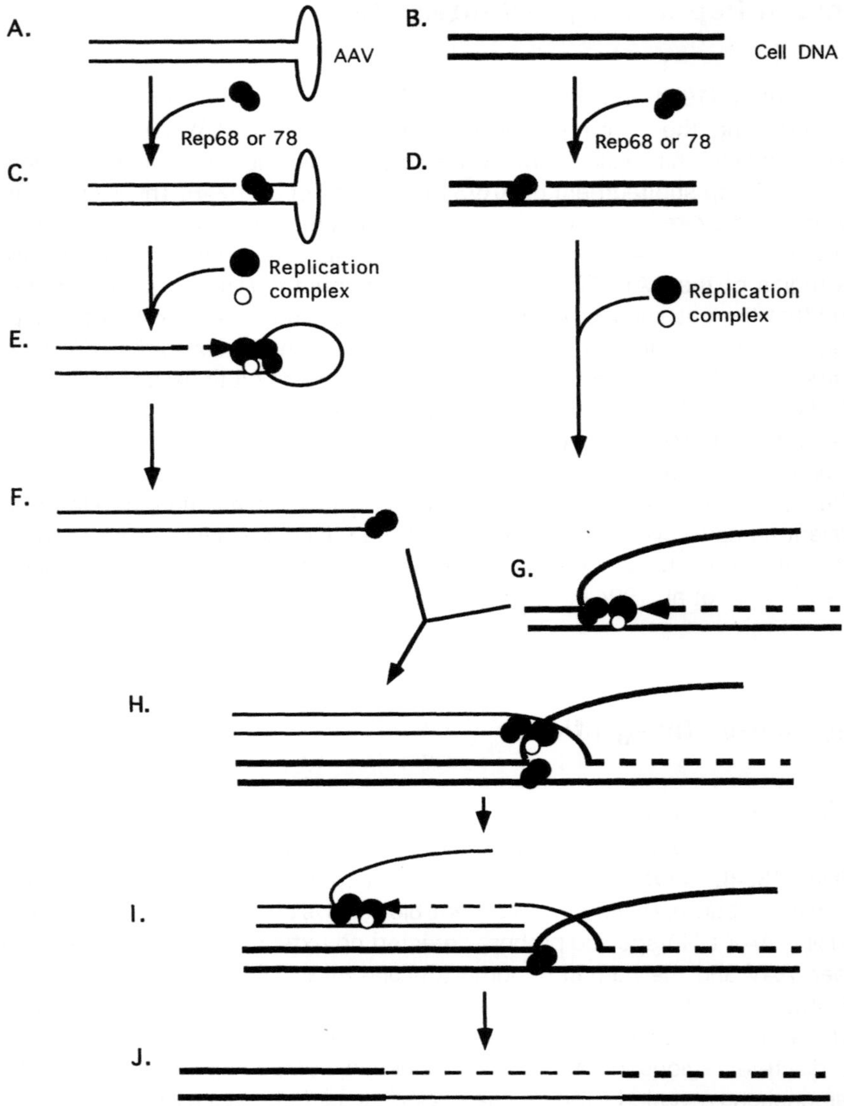

Fig. 3A–J. Model for replication-mediated recombination. **A, B** The DNA substrates for Rep68 and Rep78 binding. The Inverted terminal repeat (itr) of adeno-associated virus (AAV) and the integration site on chromosome 19, respectively. **C, D** Rep endonuclease activity on the viral and cellular substrates. Rep remains covalently attached to the 5′ end at the nick site. **E** Association of a cellular replication complex with Rep at the nicked site. DNA synthesis initiating at the free 3′-OH at the nick site with Rep functioning as a helicase for lead strand synthesis through the itr. **F** Extended itr with Rep covalently attached at the 3′ end of the duplex DNA. **G** Unidirectional leading strand synthesis of the linear duplex DNA within the integration locus. **H** Complex formed between the replicating human DNA and the extended AAV itr. Such a complex could be mediated by protein-protein (rep-Rep) interactions or through tertiary DNA-Rep-DNA interactions. **I** Strand switch of the replication complex produces the nonhomologous recombination junction. **J** Products of the replication/integration reaction. (From Kotin 1994)

Rep involvement. Recent data describe the organization of proviral DNA in cell lines produced by transfection of either *rep* plus or *rep* minus AAV *neo* plasmids: *rep* plus plasmids only integrated into AAVS1 (Shelling and Smith 1994). The *cap* minus-*rep* plus plasmids integrated specifically; thus the AAV structural proteins were not essential for targeted integration (Shelling and Smith 1994).

To elucidate the mechanism of targeted integration and determine how Rep may be involved, the cloned P1 fragment was used as a template in cell-free replication assays, using conditions established for replication of AAV (Ward and Berns 1991) and cloned, truncated ITR (Chiorini et al. 1994a). Analysis of the newly synthesized DNA products demonstrated that the cloned P1 fragment functions *in cis* as a Rep-dependent origin of replication. The labeling pattern of the replicated DNA indicated that leading strand DNA synthesis proceeded unidirectionally. Hybridization results using the radiolabeled replication products as probes against single-stranded vector targets provided additional support for asymmetric replication. Significantly, in latently infected cell lines with AAV provirus integrated into AAVS1, the proviral DNA was situated in what can now be described as downstream of the Rep binding site and TRS homolog.

Our results indicate that Rep not only binds to cellular DNA at the integration locus, but allows replication to initiate from a cloned subfragment of the integration site in a cell-free system. The in vitro replication we observed was unidirectional and highly processive. Thus we now postulate that targeted integration of AAV DNA into chromosome 19 is mediated by Rep proteins and involves limited DNA synthesis. This mechanism of integration represents a novel viral pathway for directed recombination.

References

Ashktorab H, Srivastava A (1989) Identification of nuclear proteins that specifically interact with adeno-associated virus type 2 inverted terminal repeat hairpin DNA. J Virol 63: 3034–3039

Batchu RB, Kotin RM, Hermonat PL (1994) The regulatory rep protein of adeno-associated virus binds to sequences within the c-H-ras promoter. Cancer Lett 86: 23–31

Beaton A, Palumbo P, Berns KI (1989) Expression from the adeno-associated virus p5 and p19 promoters is negatively regulated in trans by the rep protein. J Virol 63: 4450–4454

Berns KI, Kotin RM, Labow MA (1988) Regulation of adeno-associated virus DNA replication. Biochim Biophys Acta 951: 425–429

Cheung AK, Hoggan MD, Hauswirth WW, Berns KI (1980) Integration of the adeno-associated virus genome into cellular DNA in latently infected human Detroit 6 cells. J Virol 33: 739–748

Chiorini JA, Weitzman MD, Owens RA, Urcelay E, Safer B, Kotin RM (1994a) Biologically active Rep proteins of adeno-associated virus type 2 produced as fusion proteins in Escherichia coli. J Virol 68: 797–804

Chiorini JA, Wiener SM, Owens RA, Kyostio SRM, Kotin RM, Safer B (1994b) Sequence requirements for stable binding and function of adeno-associated virus 2 Rep68 protein on the itr. J Virol 68: 7448–7457

Hermonat PL (1989) The adeno-associated virus Rep78 gene inhibits cellular transformation induced by bovine papillomavirus. Virology 172: 253–261

Hermonat PL (1991) Inhibition of H-ras expression by the adeno-associated virus Rep78 transformation suppressor gene product. Cancer Res 51: 3373–3377

Hermonat PL (1992) Inhibition of bovine papillomavirus plasmid DNA replication by adeno-associated virus. Virology 198: 329–333

Hermonat PL, Labow MA, Wright R, Berns KI, Muzyczka N (1984) Genetics of adeno-associated virus: isolation and preliminary characterization of adeno-associated virus type 2 mutants. J Virol 51: 329–339

Hong G, Ward P, Berns KI (1992) In vitro replication of adeno-associated virus DNA. Proc Natl Acad Sci USA 89: 4673–4677

Im DS, Muzyczka N (1989) Factors that bind to adeno-associated virus terminal repeats. J Virol 63: 3095–3104

Im D-S, Muzyczka N (1990) The AAV origin binding protein is an ATP-dependent site-specific endonuclease with DNA helicase activity. Cell 61: 447–457

Im DS, Muzyczka N (1992) Partial purification of adeno-associated virus Rep78, Rep52, and Rep40 and their biochemical characterization. J Virol 66: 1119–1128

Katze MG, DeCorato D, Krug RM (1986) Cellular mRNA translation is blocked at both initiation and elongation after infection by influenza virus or adenovirus. J Virol 60: 1027–1039

Khleif SN, Myers T, Carter BJ, Trempe JP (1991) Inhibition of cellular transformation by the adeno-associated virus rep gene. Virology 181: 738–741

Kotin RM (1994) Prospects for the use of adeno-associated virus as a vector for human gene therapy. Hum Gene Ther 5: 793–801

Kotin RM, Siniscalco M, Samulski RJ, Zhu X, Hunter L, Laughlin CA, McLaughlin S, Muzyczka N, Rocchi M, Berns KI (1990) Site-specific integration by adeno-associated virus. Proc Nat Acad Sci USA 87: 2211–2215

Kotin RM, Menninger JC, Ward DC, Berns KI (1991) Mapping and direct visualization of a region-specific viral DNA integration site on chromosome 19q13-qter. Genomics 10: 831–834

Kotin RM, Linden RM, Berns KI (1992) Characterization of a preferred site on human chromosome 19q for integration of adeno-associated virus DNA by non-homologous recombination. EMBO J 11: 5071–5078

Krausslich HG, Nicklin MJ, Toyoda H, Ethcison D, Wimmer E (1987) Poliovirus proteinase 2A induces cleavage of eukaryotic initiation factor 4F polypeptide p220. J Virol 61: 2711–2718

Kyostio S, Owens R, Weitzman M, Antoni B, Chejanovsky N, Carter B (1994) Analysis of adeno-associated (AAV) wild-type and mutant Rep proteins for their ability to regulate negatively AAV p5 and p19 mRNA levels. J Virol 68: 2947–2957

Labow MA, Berns KI (1988) The adeno-associated virus rep gene inhibits replication of an adeno-associated virus/simian virus 40 hybrid genome in cos-7 cells. Mol Cell Biol 7: 1320–1325

Labow MA, Hermonat PL, Berns KI (1986) Positive and negative autoregulation of the adeno-associated virus type 2 genome. J Virol 60: 515–524

Labow MA, Graf LH, Berns KI (1987) Adeno-associated virus gene expression inhibits cellular transformation by heterologous genes. Mol Cell Biol 7: 1320–1325

McCarty DM, Pereira DJ, Zolotukhin I, Zhou X, Ryan JH, Muzyczka N (1994a) Identification of linear DNA sequences that specifically bind the adeno-associated virus Rep protein. J Virol 68: 4988–4997

McCarty DM, Ryan JH, Zolotukhin S, Zhou X, Muzyczka N (1994b) Interaction of the adeno-associated virus Rep protein with a sequence within the A palindrome of the viral terminal repeat. J Virol 68: 4998–5006

McLaughlin SK, Collis P, Hermonat PL, Muzyczka N (1988) Adeno-associated virus general transduction vectors: analysis of proviral structures. J Virol 62: 1963–1973

Ni T-H, Zhou X, McCarty D, Zolotukhin I, Muzyczka N (1994) In vitro replication of adeno-associated virus DNA. J Virol 68: 1128–1138

Oelze I, Rittner K, Sczakiel G (1994) Adeno-associated virus type 2 rep gene-mediated inhibition of basal expression of human immunodeficiency virus type 1 involves its negative regulatory functions. J Virol 68: 1229–1233

Owens RA, Carter BJ (1992) In vitro resolution of adeno-associated virus DNA hairpin termini by wild-type Rep protein is inhibited by a dominant-negative mutant of rep. J Virol 66: 1236–1240

Owens RA, Trempe JP, Chejanovsky N, Carter BJ (1991) Adeno-associated virus rep proteins produced in insect and mammalian expression systems: wild-type and dominant-negative mutant proteins bind to the viral replication origin. Virology 184: 14–22

Owens RA, Weitzman MD, Kyostio SR, Carter BJ (1993) Identification of a DNA-binding domain in the amino terminus of adeno-associated virus Rep proteins. J Virol 67: 997–1005

Pilder S, Moore M, Logan J, Shenk T (1986) The adenovirus E1B-55K transforming polypeptide modulates transport or cytoplasmic stabilization of viral and host cell mRNAs. Mol Cell Biol 6: 470–476

Samulski RJ, Zhu X, Xiao X, Brook JD, Housman DE, Epstein N, Hunter LA (1991) Targeted integration of adeno-associated virus (AAV) into human chromosome 19 (published erratum appears in EMBO J 1992 Mar; 11(3): 1228). EMBO J 10: 3941–3950

Shelling A, Smith M (1994) Targeted integration of transfected and infected adeno-associated virus vectors containing the neomycin resistance gene. Gene Ther 2: 1–5

Snyder R, Im D-S, Ni T, Xiao X, Samulski R, Muzyczka N (1993) Features of the adeno-associated virus origin involved in substrate recognition by the viral Rep protein. J Virol 67: 6096–6104

Srivastava A, Lusby EW, Berns KI (1983) Nucleotide sequence and organization of the adeno-associated virus 2 genome. J Virol 45: 555–564

Trempe JP, Carter BJ (1988) Regulation of adeno-associated virus gene expression in 293 cells: control of mRNA abundance and translation. J Virol 62: 68–74

Urcelay E, Ward P, Wiener SM, Safer B, Kotin RM (1995) Asymmetric replication in vitro from a human sequence element is dependent on adeno-associated virus Rep protein. J Virol 69 (in press)

Ward P, Berns KI (1991) In vitro rescue of an integrated hybrid adeno-associated virus/simian virus 40 genome. J Mol Biol 218: 791–804

Weitzman MD, Kyostio SRM, Kotin RM, Owens RA (1994) Adeno-associated virus (AAV) rep proteins mediate complex formation between AAV DNA and the human integration site. Proc Natl Acad Sci USA 91: 5808–5812

Packaging Systems for Adeno-associated Virus Vectors

J.P. TREMPE

1 Introduction

The promise and hopes that gene therapy vectors derived from adeno-associated virus (AAV) will make the leap from the laboratory bench to the clinic bedside are thoroughly described in this volume of *Current Topics of Microbiology and Immunology*. As more investigators enter this rapidly growing field, AAV's ability to deliver genes of therapeutic potential will be explored for many acquired and hereditary diseases. AAV has several characteristics that make it an appealing human gene therapy vector: (1) the virus is naturally defective in that it requires the help of a coinfecting virus, usually adenovirus, for it to efficiently complete its life cycle, (2) in the absence of helper virus infection, AAV establishes a lysogenic state, usually in a site-specific locus on chromosome 19, (3) AAV is nonpathogenic because it has not been implicated as the causative agent of any known disease, (4) AAV's small DNA genome (4680 nucleotides) allows for easy manipulation by standard recombinant DNA methodology. These characteristics have been described in other chapters of this volume but are briefly summarized here.

Adenovirus (Ad) is the most common helper for AAV although human herpesvirus type 6 (HHV-6) (THOMSON et al. 1994), herpes simplex virus (HSV), and

---

Department of Biochemistry and Molecular Biology, Medical College of Ohio, P.O. Box 10008, Toledo, OH 43699, USA cytomegalovirus (CMV) will all provide helper functions (reviewed in BERNS 1990; MUZYCZKA 1992). Five regions of the Ad genome are required for a permissive AAV infection. The E1A, E1B, E2A, E4 and VA regions play various roles in increasing AAV gene expression as well as inducing the host cell to enter the DNA synthetic phase. A helper virus is not an obligate colleague for AAV permissiveness because under some experimental conditions the virus completes its life cycle in the absence of helper coinfection (SCHLEHOFER et al. 1986; YAKOBSEN et al. 1987, 1989; YALKINOGLU et al. 1988).

The ability of AAV to integrate its genome into a specific region of chromosome 19 is a unique feature among viruses (SAMULSKI 1993; KOTIN 1994). AAV's site specificity is apparently dependent upon the replication gene encoded by the virus because vectors that lack this gene lose chromosome 19 specificity and integrate into other loci as well. One of the concerns of gene therapy vectors based on retroviruses is that one cannot predict where in the genome the provirus will integrate. Provirus integration could disrupt the host cell's normal pattern of gene expression leading to oncogenic transformation (KUNG et al. 1991). Development of AAV vectors that retain the integration capabilities of the wild-type virus is a goal in vector development and would represent a significant advance in gene therapy technology.

Although AAV seroprevalence approaches 80% for the general population (PARKS et al. 1970), the virus has not been implicated as the causative agent for any known disease. There is also speculation that the virus could provide a protective effect against some cancers (MAYOR et al. 1976; GEORG-FRIES et al. 1984). Nonpathogenicity is an asset for a viral gene delivery vector. But the long-held concept that AAV is harmless has been challenged by the detection of AAV-2 DNA in normal and diseased human uterine tissue and material from first trimester spontaneous abortions (TOBIASCH et al. 1994). Nested polymerase chain reaction (PCR) amplifications were used to detect AAV-2 DNA in biopsy material from cervix uteri. Larger amounts of AAV-2 DNA were detected using Southern hybridizations of DNA from spontaneous abortion material. In situ hybridization analysis localized the AAV-2 DNA in the villous fraction (trophoblast) of the placenta but not in the embryo or decidua. The presence of AAV-2 DNA in normal uterine tissues suggests that AAV may have established latency at this site. Therefore, before determining whether AAV plays a pathological role in first trimester abortions, normal placental villi should be investigated for AAV replication (TOBIASCH et al. 1994). It is unlikely that this recent observation will alter the course of AAV as a gene therapy vector because precautions are taken during vector preparation to eliminate wild-type AAV contamination. Nevertheless, these data emphasize the importance of developing safe and effective means of vector preparation.

Two significant challenges currently prevent AAV from vaulting to the forefront of viral gene delivery systems. The first of these is the development of AAV vectors that will integrate into the chromosome 19 site. The importance of this challenge is described above and is dealt with elsewhere in this volume. The second major challenge is the development of an efficient packaging system that will enable cost-effective production of high titer recombinant vectors. The widespread development and use of AAV vectors is also hindered by cumbersome vector production and purification procedures. The remainder of this chapter is devoted to past, current and future efforts to develop efficient AAV packaging systems.

2 Molecular Biology of Adeno-associated Virus

AAV is a single-stranded DNA virus. Plus and minus strands of DNA are packaged into virions and are equally infectious (Berns 1990; Muzyczka 1992). At each end of the 4680 nucleotide genome are 145 base terminal repeat sequences. In the single-stranded form, the terminal repeat sequence snaps back on itself to form a T-shaped three-way junction. This structure serves as the viral origin of DNA replication from which a large pool of replicative form, double-stranded DNA arises during AAV and Ad coinfection. The 145 base terminal repeat sequences are the only known *cis*-acting elements that are required for packaging of AAV DNA. Vectors containing heterologous genes flanked by only the terminal 145 bases of AAV DNA are packaged into infectious particles (Flotte et al. 1993). The terminal repeat sequence also contains elements that function as a transcription promoter. Using only the terminal repeat sequence as a promoter, Flotte et al. (1993) demonstrated efficient expression of a *cat* reporter gene or a truncated cystic fibrosis cDNA in several different cell lines. Although there are no canonical TATA boxes, the terminal repeats contain several Sp1 sites and strong homology to the consensus *Inr* site found in many genes (cited in Flotte et al. 1993).

DNA sequence analysis and extensive mutagenesis of AAV has revealed that between the terminal repeat sequences are two large translation open reading frames (Srivastava et al. 1983; Hermonat et al. 1984; Tratschin et al. 1984a). The reading frame on the right side of the genome contains the capsid (*cap*) gene which encodes the three structural proteins, VP-1, -2 and -3 (Fig. 1). These proteins make up the virion and are translated from two differentially spliced mRNAs that originate from the transcription promoter at map unit 40, p40. On the left side of the genome is the replication (*rep*) gene which encodes at least four replication (Rep) proteins. The two larger proteins, Rep78 and Rep68, are translated from unspliced and spliced mRNAs that originate from a transcription promoter at map unit 5, p5. Rep52 and Rep40 are translated from similarly spliced mRNAs that come from a third promoter at map unit 19, p19. The Rep proteins are pleiotropic effectors of multiple functions involved in the AAV life cycle (Berns 1990; Muzyczka 1992). The Rep78 and Rep68 proteins are essential for viral DNA replication. These proteins also positively and negatively regulate viral gene expression (Tratschin et al. 1986; Beaton et al. 1989; Kyöstiö et al. 1994). Rep78 and Rep68 also inhibit gene expression from a variety of heterologous transcription promoters (Labow et al. 1987; Trempe and Carter

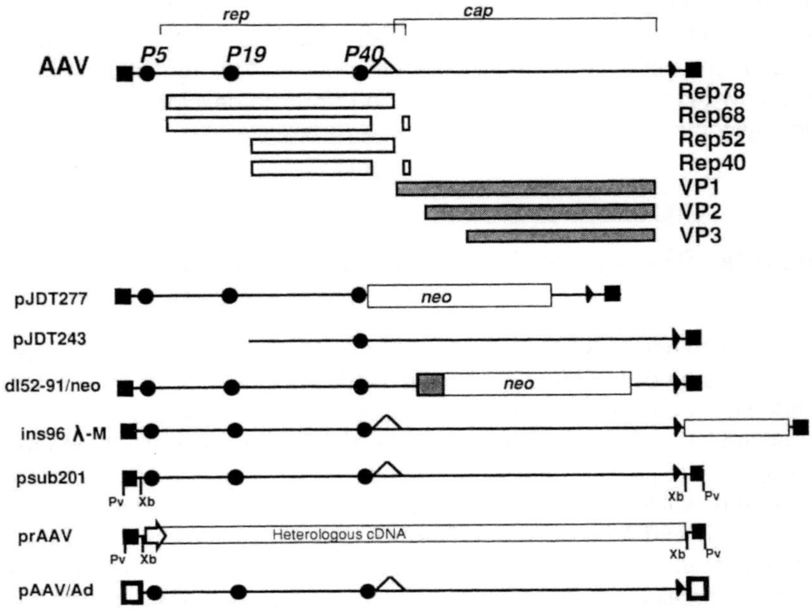

Fig. 1. Adeno-associated Virus (AAV) genome and plasmids used in vector packaging. The AAV genome is indicated at the *top* of the figure. The locations of the replication (*rep*) and capsid (*cap*) genes are indicated by the *brackets*. The *filled boxes* on each end of the genome are the terminal repeat sequences. The *filled circles* are the locations of the transcription promoters. The single poly-adenylation site is denoted by the *small arrowhead* at the right end of the genome. The *caret* next to the p40 promoter is the location of the splice site. The Rep proteins are the *open rectangles* and the *hatched rectangles* are the Cap proteins. The pJDT243 plasmid was used to package pJDT277 vector sequences. The pJDT277 plasmid contains the bacterial *neo* resistance gene in place of the *cap* gene. The dl52–91/neo plasmid contains the *neo* gene under the control of the SV40 early promoter which is indicated by the *shaded square* at the left end of the *neo* gene. The ins96λ-M plasmid, which was used to package dl52–91/neo, contains the complete AAV genome with a 1kbp DNA insert from bacteriophage λ at map unit 96. The psub201 plasmid contains the AAV genome with synthetic *Xba*I linkers located 191 bp from the ends of the genome. The terminal sequences in psub201 lack the terminal 15 bp of wild-type AAV and the AAV sequences are joined to the plasmid via *Pvu*II sites. The prAAV plasmid is an example of a recombinant vector derived from psub201. The *arrow* indicates a heterologous transcription promoter. The *open rectangles* at the ends of pAAV/Ad are the terminal 107 bp from Ad

1988; ANTONI et al. 1991; HERMONAT 1991). The Rep proteins limit cell proliferation when introduced into cell culture in the absence of Ad coinfection (LABOW et al. 1987; MENDELSON et al. 1988b; HERMONAT 1989; KHLEIF et al. 1991). The smaller Rep52 and Rep40 are not required for DNA replication but they apparently play a role in assembly of an infectious particle (CHEJANOVSKY and CARTER 1989).

Other elements in the AAV genome include a single poly-adenylation site located at map unit 96 and various smaller open reading frames. Two proteins encoded in the intron sequences downstream from p40 have been detected in in vitro translation assays (BECERRA et al. 1988). The significance of these open reading frames has not been determined.

3 Production of Recombinant Adeno-associated Virus Vectors

3.1 General Strategy

Four components are required for packaging an AAV vector: tissue culture cells, AAV replication and capsid proteins, the vector DNA, and the helper virus (Fig. 2). The AAV replication and capsid proteins are provided in *trans*, usually by a plasmid carrying these genes. The vector DNA plasmid contains the 145 bp AAV terminal repeat sequences and a transcription cassette. The transcription cassette contains a heterologous transcription promoter linked to the gene of interest. Several laboratories (MILLER et al. 1994; GOODMAN et al. 1994) have attained gene expression mediated by cell type-specific enhancer and promoter elements carried in an AAV vector. Our laboratory has also used the surfactant protein B transcription promoter (BOHINSKI et al. 1993) in an AAV vector to drive marker gene expression in transformed lung cell lines (Xing and Trempe, unpublished results). The third element required in packaging is the helper virus and the fourth element is the cell line. Various cell lines have been used successfully for this process including 293 cells (GRAHAM et al. 1977), HeLa, and KB cells. The 293 cells may be the preferred cell line because of their high efficiency of transfection (ALWINE 1985).

The basic design of an AAV vector packaging scheme is outlined in Fig. 2. The vector plasmid (prAAV) and the plasmid bearing the *rep* and *cap* genes (pAAV/Ad) are cotransfected onto Ad-infected tissue culture cells. In the presence of Ad, the Rep proteins rescue and amplify the recombinant AAV (rAAV) genome from the vector plasmid yielding a large pool of replicative form DNA. The Cap proteins sequester single-stranded rAAV genomes into a preformed virion thus producing an infectious particle. When the culture shows a significant cytopathic effect from Ad infection (usually within 48–72 h), the cells are harvested and concentrated by low speed centrifugation. Most of the recombinant AAV vector (rAAV) remains cell associated and there is not an appreciable loss of vector into the culture supernatant. The cell pellet is frozen and thawed three to four times and the lysate heated at 55°–60°C for 30–60 min to inactivate Ad. After a low speed centrifugation step to remove debris, the crude vector preparation is purified via one or more $CsCl_2$ equilibrium gradient centrifugations. AAV vectors with a genome equal in size to the wild-type genome will band at a density of 1.41 g/cm^3 allowing them to be separated from Ad helper whose density is approximately 1.35 g/cm^3 (DELAMAZA and CARTER 1978). The heat inactivation step is sufficient to inactivate all of the Ad that may be produced and use of crude rAAV preparation in cell culture is a common practice. If the rAAV is to be used in animal experiments, vector purification should be performed via multiple $CsCl_2$ gradient centrifugations. Even minor contaminations of heat-inactivated Ad can produce a local inflammatory response at the sight of virus inoculation.

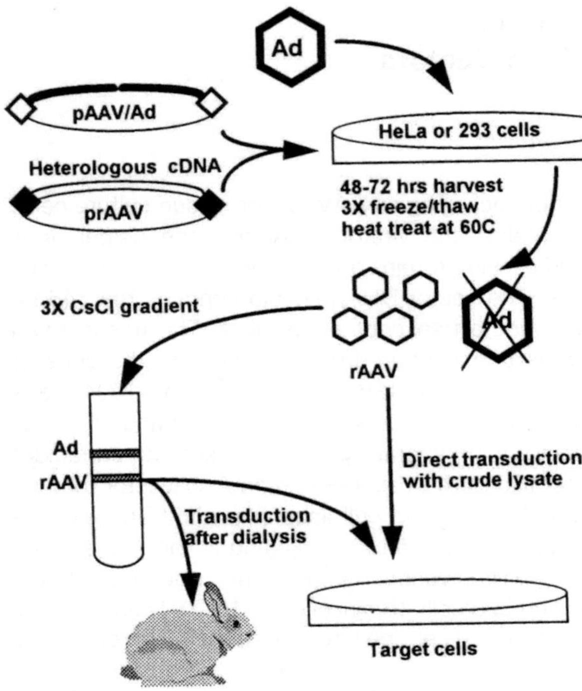

Fig. 2. Protocol for packaging recombinant adeno-associated virus(rAAV) vectors by plasmid cotransfections. Details are provided in the text

After isolation of the crude or purified rAAV, titration of the vector is performed in different ways. The single-stranded vector DNA is isolated, separated by agarose gel electrophoresis, and probed by Southern hybridization to determine the copy number of packaged genomes (SAMULSKI et al. 1989; RUSSELL et al. 1994). It has been our experience that the use of slot blot hybridizations are unacceptable for quantitating virus titers. Defective interfering genomes that are packaged into particles give an inflated signal that does not accurately represent the amount of full length, packaged genomes. By separating packaged AAV genomes on an agarose gel an estimate of the amount of full length DNA can be made. Alternatively, the rAAV titer is measured phenotypically in cell culture if the vector carries a marker gene (HERMONAT and MUZYCZKA 1984; TRATSCHIN et al. 1985). The latter method of titration yields a titer in infectious units (IUs) whereas the former yields a total particle number. The two values are somewhat interconvertable because one IU of wild-type AAV equals 40–100 particles (CARTER et al. 1979). An IU is defined as the number of particles required to infect one cell and express a phenotype. Since AAV does not form plaques, AAV *cap* gene expression in the nucleus of an AAV and Ad coinfected cell is the phenotype that is determined for the wild-type virus. For an AAV vector carrying a *neo* resistance gene, the measurable phenotype would be the conversion of cells to resistance to the drug G418 (Hermonat and Muzyczka 1984; Tratschin et al. 1985; McLaughlin et al. 1988).

3.2 Packaging via Plasmid Transfections

Early work with AAV-based vectors used bacterial drug resistance marker genes inserted in place of the *rep* or *cap* gene sequences (Hermonat and Muzyczka 1984; Tratschin et al. 1984b, 1985; McLaughlin et al. 1988). In order to package the vector genome, two different plasmids were transfected onto Ad-infected cells. Tratschin et al. (1985) cotransfected a vector-bearing plasmid (pJDT277, Fig. 1) with a second plasmid containing either the wild-type AAV genome or a plasmid that expressed only the *cap* gene. Hermonat and Muzyczka (1984), and McLaughlin et al. (1988) cotransfected a vector plasmid (dl52-91/*neo*) with a second plasmid that contained the entire AAV genome with a 1 kbp insert of bacteriophage λ DNA (ins96λ-M). The AAV genome with the λ DNA insert is rescued from the plasmid backbone and amplified in the cell, but it is too large to assemble into an infectious particle. In all of these preparations the rAAV was contaminated with wild-type virus at levels ranging from 10%–50%. Recombination between two cotransfected plasmids that shared significant sequence identity, or loss of the λ DNA fragment, is responsible for generation of wild-type virus. After the initial plasmid cotransfection, the resulting rAAV could be further amplified by subsequent passage in suspension KB cells that were coinfected with Ad (Tratschin et al. 1985). The contaminating wild-type AAV generated in the initial transfection produced Rep and Cap proteins that allowed for further amplification of the rAAV vector. Tratschin et al. (1985) used a vector genome that was smaller than wild-type length which resulted in a lower density particle (1.36–1.39 g/cm^3) that could be separated from wild-type virus by banding in $CsCl_2$ equilibrium density gradients. Unfortunately, these preparations still had significant wild-type virus contamination. Other investigators used similar approaches that yielded much less wild-type contamination (Lebkowski et al. 1988).

To eliminate the production of wild-type virus and increase the yield of rAAV, two plasmid constructs were developed that share no sequence identity (Fig. 1) (Samulski et al. 1987, 1989). A plasmid called psub201 was constructed containing the right terminal 191 bp at both ends of the viral coding sequences (Samulski et al. 1987). Two *Xba*I cleavage sites were located between the viral coding domains and the terminal repeat sequences. These convenient restriction endonuclease sites allow for removal of AAV coding sequences so that nonviral sequences can be inserted between the terminal repeats. To complement vectors derived from psub201, Samulski et al. (1989) constructed pAAV/Ad, which lacks the terminal 192 bp but retains the coding sequences for the *rep* and *cap* genes. In pAAV/Ad, the normal AAV terminal repeat sequences were replaced by the first 107 bp of the Ad type 5 genome. It was originally thought that the use of Ad terminals would mediate amplification of the AAV sequences in the presence of Ad helper. This did not occur, but the presence of the Ad sequences was fortuitous because they promoted a higher level of AAV *cap* gene expression than a similar plasmid that lacked Ad sequences. Using pAAV/Ad and vectors derived from psub201, rAAV stocks were prepared that had no detectable wild-type virus contamination because the two plasmids do not share any AAV sequence identity. Moreover, cotransfection of these plasmids onto Ad-infected 293 cells yielded 3×10^7 particles (or 3×10^5 IUs) from one 10 cm dish (SAMULSKI et al. 1989).

The minimal sequence required for AAV DNA replication and packaging has been refined to 145 bp (FLOTTE et al. 1993). A vector containing a truncated cystic fibrosis transmembrane conductance regulator cDNA and only 145 bp of AAV terminal repeat sequence at each end of the transcription cassette was efficiently encapsidated and used to transduce cells without selection. This observation increases the size of heterologous DNA that can be packaged into an AAV virion to approximately 4500 bases.

Numerous laboratories are now using the packaging protocol based upon cotransfection of pAAV/Ad and various vector plasmids onto Ad-infected cells. Although this technology is the method of choice for packaging rAAV, its limitations are a hindrance to further development of AAV as a human gene therapy vector. Plasmid transfections are inherently inefficient and cotransfection of two or more plasmids is even more inefficient. Although wild-type AAV can be concentrated to greater than 10^{12} particles per ml, rAAV titers have yet to reach such a prodigious amount. The reason for this discrepancy is not known. Transfected plasmids bearing the *rep* and *cap* genes are not amplified in the cell, thus the copy number of transcription templates is limited by transfection efficiency. In an AAV and Ad coinfected cell, the copy number of replicative form AAV DNA approaches 10^6/cell leading to a very high level of viral Rep and Cap proteins. Wild-type AAV genomes are always replicated and encapsidated more efficiently than rAAV genomes in the same cell (HERMONAT and MUZYCZKA 1984; MCLAUGHLIN et al. 1988; Trempe and Carter, unpublished observation). Thus there may be other unidentified *cis*-acting elements in the genome that promote replication and packaging. Another limitation of the current technology is the need for cumbersome, labor-intensive purification of rAAV over multiple $CsCl_2$ gradients.

Clearly, there is a great need for an efficient and easy method of packaging AAV vectors. Although no such system has been developed, the field is advancing at a rapid rate and such systems may soon be available to the gene therapy community. Several of the approaches that are being pursued are described in the remainder of this chapter.

3.3 Adeno-associated Virus Protein-Expressing Cell Lines

For both retrovirus and Ad expression vectors, the most efficient means of producing recombinant virions is through the use of packaging cell lines (reviewed in BERKNER 1992; MORGAN and ANDERSON 1993). The cell lines that are widely used constitutively express viral proteins that act in *trans* to amplify and package infectious viral genomes. Although efficient AAV packaging cell lines have not been developed, results from several laboratories suggest that they may soon be available. MENDELSON et al. (1988b) was the first to isolate cell lines that constitutively expressed AAV Rep proteins suggesting that the development of AAV packaging cell lines was feasible. A true packaging cell line was subsequently developed by VINCENT et al. (1990). HeLa cell lines were transfected with AAV plasmids that lacked the viral origins of replication. Several cell lines were isolated with low copy numbers of integrated AAV genomes. One of the cell lines was able to generate rAAV stocks at a titer of 10^4/ml. The low level of rAAV produced from this cell line may be due to an overall low level of Rep and Cap proteins expression. Unfortunately, these stocks were contaminated with a low level of wild-type AAV.

Efforts by several laboratories to develop cell lines that express high levels of Rep or Cap proteins were unsuccessful. This was due primarily to the well known, but not well understood, antiproliferative effects of the AAV *rep* gene. As mentioned earlier, the Rep78 and Rep68 proteins suppress gene expression from both cellular and viral transcription promoters in transient assays. In long term assays the Rep proteins prevent the growth of cells in culture. The mechanism of this inhibition has not been well defined and various hypotheses have been tested. One potential mechanism suggests that inhibition of mRNA translation may suppress gene expression (TREMPE and CARTER 1988). Another mechanism might involve inhibition of gene expression at the transcriptional level (BEATON et al. 1989; ANTONI et al. 1991; HERMONAT 1994; BATCHU et al. 1994). A third possibility is that expression of the Rep78 protein inhibits cellular DNA replication thus limiting cell proliferation (BANTEL-SCHAAL and ZUR HAUSEN 1988; HEILBRONN et al. 1990; HERMONAT 1992; YANG et al. 1995). These possibilities are not mutually exclusive, and given the pleiotropic effects of the Rep proteins, it is likely that Rep proteins mediate their antiproliferative effects through several mechanisms.

The chances of developing an efficient packaging cell line have been improved with the isolation of 293 cell-based lines that inducibly express the AAV Rep proteins (YANG et al. 1994). The plasmid used to isolate these cell lines contained a 719 bp DNA fragment of the mouse metallothionein I transcription promoter attached to the *rep* gene from nucleotide 263–2233 of the AAV genome. Initially only one cell line, Neo6, was isolated. But several subclones were subsequently isolated that displayed different growth characteristics and levels of Rep78 expression. Neo6 and Neo40 (one of the subclones) expressed Rep78 and a trace of Rep68 upon induction by heavy metal salts in the growth medium. The level of Rep78 was 50%–75% of the amount of Rep78 produced from a similar number of AAV and Ad coinfected 293 cells. Rep52 was expressed constitutively at a low level and was only slightly induced upon heavy metal induction and Ad infection. Rep40 expression has not been detected in Neo6 or Neo40 cells.

The Rep78 produced by Neo6 and Neo40 cell lines is fully functional for supporting AAV DNA replication and *trans*-activating *cap* gene expression. Transfection of a plasmid bearing a mutant AAV genome containing a Gly for Met change in the initiation codon for the Rep52 protein was amplified and packaged into an infectious particle (YANG et al. 1994; CHEJANOVSKY and CARTER 1989). This mutant genome cannot normally form an infectious particle because of an unknown defect in the accumulation of single-stranded DNA. Therefore the Rep52 produced in the Neo6 cells is probably functional. We were also able to package a rAAV that contains only a *cat* reporter gene by cotransfecting a plasmid that expresses the AAV *cap* gene. This is further evidence that Neo6 and Neo40 cells supply all of the known Rep protein functions.

An interesting feature of these cell lines is that, upon induction of Rep expression, cell proliferation slows dramatically. Moreover, when Rep expression is turned off, the cells recover and begin growing again. Cell cycle analysis of induced cultures revealed that there is an accumulation of cells in the S phase of the cycle. This was especially evident for the Neo40 subclone. Tritiated thymidine uptake experiments have recently indicated that the induced cultures are synthesizing about fivefold less DNA than the uninduced cultures (Chen and Trempe, unpublished observation). Although the cells appear to be in S phase, they are not capable of carrying on normal DNA replication. Thus the Rep protein may block 293 cell proliferation by inhibiting cellular DNA replication. We are currently investigating the mechanism of this inhibition.

Since establishing Rep-expressing cell lines, we have attempted to introduce functional AAV *cap* genes into Neo6 cells by various means. Using various versions of *cap* genes in Epstein-Barr virus (EBV) replicon plasmids, we have been unsuccessful in obtaining a fully functional packaging cell line. When introduced into Neo6 cells in transient transfections, the EBV/*cap* gene plasmids produce VP-1, -2, and -3, but stable cell lines containing these plasmids have not been established. There is evidence that an AAV particle can exert a cytostatic effect during the G2 phase of the cell cycle (WINOCOUR et al. 1988). AAV Cap proteins can form virion-like structures in the absence of AAV single-stranded DNA (MYERS and CARTER 1980) and these virions may inhibit cell proliferation. Our results and these reports indicate that there might be some antiproliferative effects exerted on cells from high level expression of the *cap* gene.

Other Rep protein-expressing cell lines have been constructed using an inducible transcription promoter (HÖLSCHER et al. 1994). Two cell lines (HeM1, HeM2) were stably established using the glucocorticoid-responsive transcription promoter from the mouse mammary tumor virus long terminal repeat attached to the AAV *rep* gene from nucleotides 264–2260 of the AAV genome. Both lines expressed Rep78 upon dexamethasone induction of *rep* gene expression and one of them (HeM1) constitutively expressed Rep52. Both cell lines complimented a *rep* gene mutant AAV genome for DNA replication. Both cell lines *trans*-activated the AAV *cap* gene promoter yielding all three structural proteins, but neither produced infectious progeny or accumulated single-stranded DNA. If an additional plasmid that expresses any one of the four Rep proteins was introduced into the cells along with the *rep* gene mutant plasmid, infectious progeny could be produced. The authors speculate that the level of Rep expression may be too low to allow for packaging and that inclusion of additional Rep proteins ameliorates the defect. This is supported by the observation that a plasmid that expresses only Rep78 complemented the cell line and produced infectious progeny. This work also calls into question the hypothesis that Rep52 is required for accumulation of single-stranded DNA. HeM2 cells do not synthesize detectable Rep52 but when a Rep78- or Rep68-producing plasmid was introduced into HeM2 cells, infectious progeny virus was produced. One explanation for this observation is that low, undetectable levels of Rep52 are present and functional in these cells. Another explanation could be that there is functional redundancy in Rep78 and Rep68 such that if enough of them are produced, Rep52 would not be required for single-stranded DNA accumulation.

Induction of Rep78 expression had no effect on the plating efficiency of either HeM1 or HeM2. This is in stark contrast to the growth characteristics of the 293 cell-derived Neo6 and Neo40 lines (YANG et al. 1994). The only difference in the *rep* genes inserted into the respective cell lines is in the COOH-terminal end of the Rep68 protein. In the HeM1 and HeM2 lines the *rep* gene was constructed to contain the normal COOH-terminal end of the Rep68 protein (HÖLSCHER et al. 1994). By contrast, in the plasmid expression vector used in construction of the Neo6 and Neo40 cells, the normal COOH-terminal end of Rep68 is lost and it acquires an additional 36 amino acids at the COOH-terminal from sequences upstream of the AAV polyadenylation site (Trempe, unpublished observation). This point however, may be irrelevant because both types of cell line express predominantly Rep78 and little Rep68.

The differences in growth rate between the 293 and HeLa Rep-expressing cell lines may be explained in terms of cell type differences. The inhibitory effect exerted by Rep proteins in transient expression assays is stronger in 293 cells than in HeLa cells (TRATSCHIN et al. 1985, 1986). Thus HeLa cells may be able to tolerate Rep expression more readily than 293 cells.

4 Future Directions

Now that two different types of cell lines have been created that inducibly express functional Rep proteins, the logical next step is to add the *cap* genes to the cell lines. One means of adding the *cap* genes to these systems would be to construct Ad vectors that express them. Such a vector would serve a dual purpose in providing both Cap proteins and Ad helper functions. Recently, a Cap protein-expressing Ad vector has been reported (SAMULSKI et al. 1994). This Ad vector expresses high levels of Cap proteins with kinetics that are similar to the normal rate of Cap synthesis found in wild-type AAV infections. It will be very interesting to see if this chimeric Ad vector will complement a Rep-expressing cell line to amplify and encapsidate an AAV vector.

Another possible packaging system to consider is based upon the use of herpes virus as helper. Recent evidence suggests that the AAV *rep* gene homologue found in HHV-6 is functional for amplifying an AAV genome (Thomson et al. 1994). Given that herpes viruses have been used as gene delivery vectors, it might be feasible to modify HHV-6 to carry a *cap* gene. Such a vector would provide the *trans*-acting Rep and Cap proteins as well as the helper virus functions. In the Thomson et al. (1994) report, there was no quantitation of the efficiency with which HHV-6 supports AAV replication. This is a crucial question that might determine the fate of such an approach for packaging AAV vectors. If HHV-6 is an inefficient helper of wild-type AAV, its usefulness for producing high titer vector preparation would be suspect. The use of a herpes virus as a helper might be preferred over Ad because its large size and susceptibility to detergents should make it easier to separate from rAAV.

There are other systems that do not depend upon the development of packaging cell lines that could yield high levels of rAAV. One such approach was mentioned in a recent review of AAV's prospects for human gene therapy (Kotin 1994). Plasmids are available that contain the SV40 origin of replication and the AAV *rep* and *cap* genes (Labow and Berns 1988). These plasmids could be transfected into COS cells (along with a rAAV vector plasmid) and would replicate to high copy numbers due to the constitutively expressed SV40 T-antigen. In the presence of Ad, full helper function would be realized and efficient packaging may be attained. Another system that deserves consideration is the high efficiency gene transfer technique which utilizes poly-lysine DNA conjugates attached to infectious Ad (Curiel et al. 1991; Wagner et al. 1992; Fisher and Wilson 1994). In this method of gene transfer, Ad is covalently coupled to poly-lysine and then allowed to associate with plasmid DNA. The Ad/poly-lys/DNA complex is then used to infect a target cell. The Ad improves the efficiency of plasmid uptake (and gene expression) by essentially providing a "receptor-mediated" uptake of DNA. The Ad particle also protects the accompanying DNA from lysosomal degradation leading to more efficient gene transfer. Transfer of an rAAV vector plasmid and a *rep* and *cap* gene plasmid such as pAAV/Ad via an Ad poly-lysine conjugate may produce high titers of vector. This technique would be amenable to packaging rAAV vectors in suspension cultures which would facilitate vector production on a larger scale.

An idealized rAAV vector production scheme is diagrammed in Fig. 3. The heterologous vector DNA would be created on a plasmid, but after an initial packaging, vector seed stocks could be used for future amplification of the vector. The helper virus may be a defective Ad, or an Ad vector that expresses the Cap proteins, or a herpes virus. The helper virus and vector plasmid (or seed virus) would be introduced into the packaging cell lines that express the AAV Rep and Cap proteins. Packaging cells grown in suspension culture would facilitate large scale vector production. Minimal, but efficient, purification of rAAV vectors

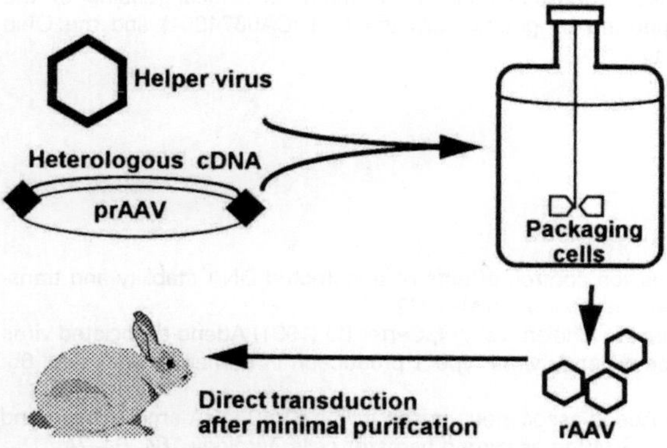

Fig. 3. An idealized packaging protocol using adeno-associated virus(AAV) packaging cells. Details are provided in the text

would eliminate equilibrium centrifugations that are currently used. Purification schemes using chromatography or differential salt fractionation may prove useful.

Finally, it is worth considering the possibility of a completely in vitro packaging system for AAV vectors. Baculovirus vectors have been developed that direct the synthesis of large amounts of the three Cap proteins that self-assemble into virus-like particles (RUFFING et al. 1992). Baculovirus and prokaryotic vectors are also available that express biologically active Rep proteins (OWENS et al. 1991; NI et al. 1994; CHIORINI et al. 1994). In vitro packaging would involve the combination of vector DNA, Rep and Cap proteins as well as any yet to be defined cellular or Ad proteins. In vitro packaging of bacteriophage λ genomes has become a routine event in the construction of genomic and cDNA libraries. The genetics of AAV and its virion structure are apparently much less sophisticated than that of bacteriophage λ, suggesting that in vitro packaging of AAV vectors is an attainable goal. Such a system would indeed be a tremendous advancement in AAV vector development.

5 Conclusions

Adeno-associated Virus was once the province of a handful of investigators who were sometimes chided by colleagues for studying a virus whose acronym really stands for "Almost A Virus". With the acceptance and advancement of human gene therapy, this handful of investigators has expanded rapidly. As the number of investigators has grown, so has our understanding of this virus and its potential as a human gene therapy vector. The problems faced in generating high titer AAV vectors are viewed by the AAV community as surmountable technical difficulties that will be overcome in the not too distant future.

Acknowledgements. The author thanks R. Collaco and J.D. Dignam for critical reading of the manuscript. The author's work is supported by grants from the NIH (CA5874301) and the Ohio Chapter of the American Cancer Society.

References

Alwine JC (1985) Transient gene expression control: effects of transfected DNA stability and trans-activation by viral early proteins. Mol Cell Biol 5: 1034–1042

Antoni BA, Rabson AB, Miller IL, Trempe JP, Chejanovsky N, Carter BJ (1991) Adeno-associated virus rep protein inhibits human immunodeficiency virus type 1 production in human cells. J Virol 65: 396–404

Bantel-Schaal U, zur Hausen H (1988) Adeno-associated viruses inhibit SV40 DNA amplification and replication of herpes simplex virus in SV40-transformed hamster cells. Virology 164: 64–74

Batchu RB, Kotin RM, Hermonat PL (1994) The regulatory rep protein of adeno-associated virus binds to sequences within the c-H-ras promoter. Cancer Lett 86: 23–31

Beaton A, Palumbo P, Berns KI (1989) Expression of the AAV p5 and p19 promoters is negatively regulated in trans by the rep protein. J Virol 63: 4450–4454

Becerra SP, Koczot F, Fabisch P, Rose JA (1988) Synthesis of adeno-associated virus structural proteins requires both alternative mRNA splicing and alternative initiations from a single transcript. J Virol 62: 2745–2754

Berkner KL (1992) Expression of heterologous sequences in adenoviral vectors. In: Muzyczka N (ed) Viral expression vectors. Springer, Berlin Heidelberg New York, pp 39–66 (Current topics in microbiology and immunology, vol 158)

Berns KI (1990) Parvoviridae and their replication: In: Fields BN, Knipe DM (eds) Virology, vol 2, 2nd edn. Raven, New York, pp 1743–1763

Bohinski RJ, Huffman JA, Whitsett JA, Lattier DL (1993) Cis-active elements controlling lung-specific expression of human pulmonary surfactant protein B genes. J Biol Chem 268: 11160–11166

Carter BJ, Laughlin CA, delaMaza LM, Myers M (1979) Adeno-associated virus autointerference. Virology 92: 449–462

Chejanovsky N, Carter BJ (1989) Mutagenesis of an AUG codon in the adeno-associated virus rep gene: effects on viral DNA replication. Virology 173: 120–128

Chiorini JA, Weitzman MD, Owens RA, Urcelay E, Safer B, Kotin RM (1994) Biologically active rep proteins of adeno-associated virus type 2 produced as fusion proteins in Escherichia coli. J Virol 68: 797–804

Curiel DT, Agarwal S, Wagner E, Cotten M (1991) Adenovirus enhancement of transferrin-polylysine-mediated gene delivery. Proc Natl Acad Sci USA 88: 8850–8854

delaMaza LM, Carter BJ (1978) DNA structure of incomplete adeno-associated virus particles. In: Ward DC, Tattersall P (eds) Replication of mammalian parvoviruses. Cold Spring Harbor Laboratory, Cold Spring Harbor, New York, pp 193–204

Fisher KJ, Wilson JM (1994) Biochemical and functional analysis of an adenovirus-based ligand complex for gene transfer. Biochem J 299: 49–58

Flotte TR, Afione SA, Solow R, Drumm ML, Markakis D, Guggino WB, Zeitlin PL, Carter BJ (1993) Expression of the cystic fibrosis transmembrane conductance regulator from a novel adeno-associated virus promoter. J Biol Chem 268: 3781–3790

Georg-Fries B, Biederlack S, Wolf J, zur Hausen H (1984) Analysis of helper dependence, and seroepidemiology of a new human parvovirus. Virology 134: 64–71

Goodman S, Xiao X, Donahue RE, Moulton A, Miller J, Walsh C, Young NS, Samulski RJ, Nienhuis AW (1994) Recombinant adeno-associated virus mediated gene transfer into hematopoietic progenitor cells. Blood 84: 1492–1500

Graham FL, Smiley J, Russell WC, Naiva R (1977) Characteristics of a human cell line transformed by DNA from adenovirus type 5. J Gen Virol 36: 59–72

Heibronn R, Bürkle A, Stephan S, and zur Hausen (1990) The adeno-associated virus rep gene suppresses herpes simplex virus-induced DNA amplification. J Virol 64: 3012–3018

Hermonat PL (1989) The adeno-associated virus rep78 gene inhibits cellular transformation by bovine papillomavirus. Virology 172: 253–261

Hermonat PL (1991) Inhibition of H-ras expression by the adeno-associated virus rep78 transformation suppressor gene product. Cancer Res 51: 3373–3377
Hermonat PL (1992) Inhibition of bovine papillomavirus plasmid DNA replication by adeno-associated virus. Virology 189: 329–333
Hermonat PL (1994) Down-regulation of the human c-fos and c-myc proto-oncogene promoters by adeno-associated virus Rep78. Cancer Lett 81: 129–136
Hermonat PL, Muzyczka N (1984) Use of adeno-associated virus as a mammalian DNA cloning vector: transduction of neomycin resistance into mammalian tissue culture cells. Proc Natl Acad Sci USA 81: 6466–6470
Hermonat PL, Labow MA, Wright R, Berns KI, Muzyczka N (1984) Genetics of adeno-associated virus: isolation and preliminary characterization of mutants of adeno-associated virus type 2. J Virol 51: 329–339
Hölscher C, Horker M, Kleinschmidt JA, Zentgraf H, Burkle A, Heilbronn R (1994) Cell lines inducibly expressing the adeno-associated virus (AAV) rep gene: requirements for productive replication of rep-negative AAV mutants. J Virol 68: 7169–7177
Khleif SN, Myers T, Carter BJ, Trempe JP (1991) Inhibition of cellular transformation by the adeno-associated virus rep gene. Virology 181: 738–741
Kotin RM (1994) Prospects for the use of adeno-associated virus as a vector for human gene therapy. Hum Gene Ther 5: 793–801
Kung H-J Boerkoel, Carter TH (1991) Retroviral mutagenesis of cellular oncogenes: a review with insights into the mechanisms of insertional activation. In: Kung HJ, Vogt PK (eds) Retroviral insertional mutagenesis and oncogene activation. Springer, Berlin Heidelberg New York, pp 1–24 (Current topics in microbiology and immunology, vol 171)
Kyöstiö SR, Owens RA, Weitzman MD, Antoni BA, Chejanovsky N, Carter BJ (1994) Analysis of adeno-associated virus (AAV) wild-type and mutant Rep proteins for their abilities to negatively regulate AAV p5 and p19 mRNA levels. J Virol 68: 2947–2957
Labow MA, Graf LH, Berns KI (1987) Adeno-associated virus gene expression inhibits cellular transformation by heterologous genes. Mol Cell Biol 7: 1320–1325
Labow MA, Berns KI (1988) The adeno-associated virus rep gene inhibits replication of an adeno-associated virus/simian virus 40 hybrid genome in cos-7 cells. Mol Cell Biol 7: 1320–1325
Lebkowski JS, McNally MA, Okarma TB, Lerch LB (1988) Adeno-associated virus: A vector system for efficient introduction and integration of DNA into a variety of mammalian cells. Mol Cell Biol 8: 3988–3996
Mayor HD, Drake S, Stahmann J, Mumford DM (1976) Antibodies to adeno-associated satellite virus and herpes simplex in sera from cancer patients and normal adults. Am J Obstet Gynecol 126: 100–104
McLaughlin SK, Collis P, Hermonat PL, Muzyczka N (1988) Adeno-associated virus general transduction vectors: analysis of proviral structures. J Virol 62: 1963–1973
Mendelson E, Miller IL, Carter BJ (1988a) Expression and rescue of a nonselected marker from an integrated AAV vector. Virology 166: 154–165
Mendelson E, Smith MG, Miller IL, Carter BJ (1988b) Effect of a viral rep gene on transformation of cells by an adeno-associated virus vector. Virology 166: 612–615
Miller JL, Donahue RE, Sellers SE, Samulski RJ, Young NS, Nienhuis AW (1994) Recombinant adeno-associated virus (rAAV)-mediated expression of a human g-globin gene in human progenitor-derived erythroid cells. Pro Natl Acad Sci USA 91: 10183–10187
Morgan RA, Anderson WF (1993) Human gene therapy. Annu Rev Biochem 63: 191–217
Muzyczka N (1992) Use of adeno-associated virus as a general transduction vector for mammalian cells. In: Muzyczka N (ed) Viral expressions vectors. Springer, Berlin Heidelberg New York, pp 97–129 (Current topics in microbiology and immunology, vol 158)
Myers MW, Carter BJ (1980) Assembly of adeno-associated virus. Virology 102: 71–82
Ni T-H, Zhou X, McCarty DM, Zolotukhin I, Muzyczka N (1994) In vitro replication of adeno-associated virus DNA. J Virol 68: 1128–1138
Owens RA, Trempe JP, Chejanovsky N, Carter BJ (1991) Adeno-associated virus rep proteins produced in insect and mammalian expression systems: wild-type and dominant-negative mutant proteins bind to the viral replication origin. Virology 184: 14–22
Parks WP, Boucher DW, Melnick JL, Taber LH, Yow MD (1970) Seroepidemiological and ecological studies of the adenovirus-associated satellite viruses. Infect Immun 2: 716–722
Ruffing M, Zentgraf H, Kleinschmidt JA (1992) Assembly of virus-like particles by recombinant structural proteins of adeno-associated virus type 2 in insect cells. J Virol 66: 6922–6930

Russell DW, Miller AD, Alexander IE (1994) Adeno-associated virus vectors preferentially transduce cells in S-phase. Proc Natl Acad Sci USA 91: 8915–8919
Samulski RJ (1993) Adeno-associated virus: integration at a specific chromosomal locus. Curr Opin Genet Devel 3: 74–80
Samulski RJ, Chang L-S, Shenk T (1987) A recombinant plasmid from which an infectious adeno-associated virus genome can be excised in vitro and its use to study viral replication. J Virol 61: 3096–3011
Samulski RJ, Chang L-S, Shenk T (1989) Helper-free stocks of recombinant adeno-associated viruses: normal integration does not require viral gene expression. J Virol 63: 3822–3828
Samulski RJ, Xiao X, Kaplitt M, Leone Pfaff P, O'Malley K, During M (1994) Generation of AAV vectors for transduction and expression in neuronal cells in vivo. Gene therapy 21–25 Sept 1994, Cold Spring Harbor Laboratory, Cold Spring Harbor, New York (abstract)
Schlehofer JR, Ehrbar M, zur Hausen H (1986) Vaccinia virus, herpes simplex virus, and carcinogens induce DNA amplification in a human cell line and support replication of a helper virus dependent parvovirus. Virology 152: 110–117
Srivastava A, Lusby EW, Berns KI (1983) Nucleotide sequence and organization of the adeno-associated virus 2 genome. J Virol 45: 555–564
Thomson BJ, Weindler FW, Gray D, Schwaab V, Heilbronn R (1994) Human herpesvirus 6 (HHV-6) is a helper virus for adeno-associated virus type 2 (AAV-2) and the AAV-2 rep gene homologue in HHV-6 can mediate AAV-2 DNA replication and regulate gene expression. Virology 204: 304–311
Tobiasch E, Rabreau M, Geletneky K, Larue-Charlus S, Severin F, Becker N, Schlehofer JR (1994) Detection of adeno-associated virus DNA in human genital tissue and in material from spontaneous abortion. J Med Virol 44: 215–222
Tratschin J-D, Miller IL, Carter BJ (1984a) Genetic analysis of adeno-associated virus: Property of deletion mutants constructed in vitro and evidence for an adeno-associated virus replication function. J Virol 51: 611–619
Tratschin J-D, West MH, Sandbank T, Carter BJ (1984b) A human parvovirus, adeno-associated virus, as a eukaryotic vector: transient expression and encapsidation of the prokaryotic gene for chloramphenicol acetyltransferase. Mol Cell Biol 4: 2072–2081
Tratschin J-D, Miller IL, Smith MG, Carter BJ (1985) Adeno-associated virus vector for high-frequency integration, expression, and rescue of genes in mammalian cells. Mol Cell Biol 5: 3251–3260
Tratschin J-D, Tal J, Carter BJ (1986) Negative and positive regulation in *trans* of gene expression from adeno-associated virus vectors in mammalian cells by a viral *rep* gene product. Mol Cell Biol 6: 2884–2894
Trempe JP, Carter BJ (1988) Regulation of adeno-associated virus gene expression in 293 cells: Control of mRNA abundance and translation. J Virol 62: 68–74
Vincent KA, Moore GK, Haigwood NI (1990) Replication and packaging of HIV envelope genes in a novel adeno-associated virus vector system. Vaccines 90: 350–353
Wagner E, Zatloukal K, Cotten M, Kirlappos H, Mechtler K, Curiel DT, Birnstiel M (1992) Coupling of adenovirus to transferrin-polylysine/DNA complexes greatly enhances receptor-mediated gene delivery and expression of transfected genes. Proc Natl Acad Sci USA 89: 6099–6103
Winocour E, Callaham MF, Huberman E (1988) Perturbation of the cell cycle by adeno-associated virus. Virology 167: 393–399
Yakobson B, Koch T, Winocour E (1987) Replication of adeno-associated virus in synchronized cells without the addition of a helper virus. J Virol 61: 972–981
Yakobson B, Hrynko T, Peak MJ, Winocour E (1989) Replication of adeno-associated virus in cells irradiated with UV light at 254 nm. J Virol 63: 1023–1230
Yalkinoglu AO, Heilbronn R, Burkle A, Schlehofer JR, zur Hausen H (1988) DNA amplification of adeno-associated virus as a response to cellular genotoxic stress. Cancer Res 48: 3123–3125
Yang Q, Chen F, Trempe JP (1994) Characterization of cell lines that inducibly express the adeno-associated virus Rep proteins. J Virol 68: 4847–4856
Yang Q, Chen F, Ross J, Trempe JP (1995) Inhibition of Cellular and SV40 DNA Replication by the adeno-associated Virus Rep Proteins. Virology (in press)

The Challenges of Recombinant Adeno-associated Virus Manufacturing: Alternative Use of Adeno-associated Virus Plasmid/Liposome Complexes for Gene Therapy Applications

J.S. Lebkowski, T.B. Okarma, and R. Philip

1 Introduction

Recombinant adeno-associated viruses (rAAVs) have strong potential as safe and efficient transducing vectors for a variety of gene therapy applications. Wild-type AAV can infect a wide range of mammalian cell types, and, in the absence of adenovirus or herpes virus, integrates its genome into that of the host (Berns and Bohensky 1987). One major obstacle to the widespread use of rAAV is the lack of simple methods to efficiently manufacture purified recombinant virus at high titer (Rolling and Samulski 1995). Several improvements have been described (Flotte et al. 1995; Lebkowski et al. 1988; Mamounas et al. 1995), however, current methods still involve multiple, variable, time consuming procedures which result in low yields of rAAV and are not suitable for scale up for widespread application.

Several elements are prerequisites for the replication and encapsidation of rAAV vectors. A producer line infected with a species-matched adenovirus or herpes virus is necessary to create a permissive environment for rAAV replication (Fig. 1). The exact functions of these helper adenoviruses or herpes viruses which complement AAV replication are only sketchily defined to date. However, an engineered cell line which constitutively or inducibly expresses all of the adenovirus or herpes virus functions necessary to support rAAV replication could alternatively be constructed to manufacture rAAV. Such an approach would, in

Applied Immune Sciences, Inc., 5301 Patrick Henry Drive, Santa Clara, CA 95054–1114, USA

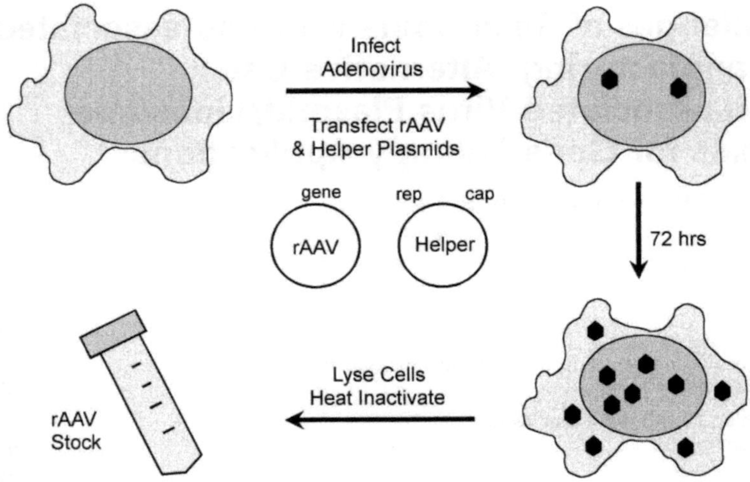

Fig. 1. Conventional production of recombinant adeno-associated virus (rAAV) vectors. A human cell line is infected with adenovirus or herpes virus and subsequently transfected with two plasmids. The first carries a rAAV genome (rAAV) which is to be amplified and packaged into virions. The second plasmid carries the wild-type AAV helper functions, *rep* and *cap*, which are required in *trans* for the replication and encapsidation of the recombinant AAV genomes into virus particles. Some 72 h after adenoviral infection and plasmid transfections, cell lysates are collected and usually heat inactivated to disable the contaminating adenovirus. The rAAV stocks may subsequently be purified using numerous protocols

theory, eliminate wild-type adenovirus or herpes virus contamination of rAAV stocks. In addition to a permissive producer cell line, rAAV production requires the presence of the wild-type AAV Rep and capsid functions. The Rep proteins catalyze rAAV genome replication while the capsid functions enable viral particle formation (BERNS and BOHENSKY 1987). Lastly, to manufacture rAAV, templates of the rAAV genome destined for packaging must be provided to the producer cell line. Traditionally, these last two sets of required elements have been provided by co-transfecting plasmids containing the rAAV genome and the *rep* and *cap* genes into the producer cell lines (Fig. 1). Alternatively, some packaging lines which constitutively or inducibly express the *rep* and *cap* genes have been engineered, but to date, have yielded low titers of recombinant virus (YANG et al. 1994; HOLSCHER et al. 1994; MUSYCKA 1992). When all of the required elements are present in the producer cells, replication of the rAAV vector ensues. Some 48–72 h later, the producer cells are lysed and rAAV is harvested (Fig. 1). Usually the contaminating helper adenovirus or herpes virus is heat inactivated or depleted using chromatography or gradient separation. Such procedures generally take 1–2 weeks to perform, yielding up to 10^{10} transducing rAAV vectors.

In this review, we describe some of our work to simplify the manufacture of rAAV virus. In our approach, rAAV genomes have been incorporated into Epstein-Barr virus (EBV) plasmids which are stably maintained as episomes in producer cell lines. To produce recombinant virus, these cell lines are infected with adenovirus and transfected with the wild-type AAV *rep* and *cap* gene functions. In additional studies, we have found that plasmids containing the inverted terminal repeats (ITRs) of AAV have unique properties that make them suitable for a variety of gene therapy applications without the requirement of viral packaging. These plasmids can be delivered to cells using numerous modes of transfection and produce higher and longer gene expression than equivalent plasmids lacking the AAV ITRs (PHILIP et al. 1994). Our investigations using these plasmids for transfection of primary cell types for clinical application are detailed.

2 Incorporation of Recombinant Adeno-associated Virus Genomes into Epstein-Barr Virus Plasmids

To permanently incorporate rAAV genomes at high copy number in candidate producer cell lines, mammalian transgene expression cassettes surrounded by the ITRs of AAV were cloned into EBV-based plasmids (Fig. 2). These plasmids contain the latent origin of replication of EBV along with the EBV nuclear antigen 1 gene and the hygromycin B resistance gene to permit mammalian cell selection (YATES et al. 1985; SUGDEN et al. 1985). When transfected into human cells, EBV-based plasmids replicate as episomes at 1–100 copies per cell. EBV plasmids containing rAAV vector genomes also remain as plasmids when

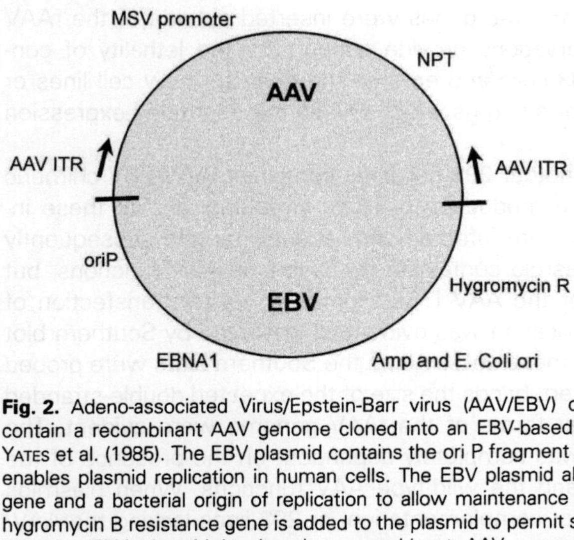

Fig. 2. Adeno-associated Virus/Epstein-Barr virus (AAV/EBV) chimeric plasmids. These plasmids contain a recombinant AAV genome cloned into an EBV-based plasmid such as that described by YATES et al. (1985). The EBV plasmid contains the ori P fragment of EBV and the EBNA 1 gene which enables plasmid replication in human cells. The EBV plasmid also contain the ampicillin resistance gene and a bacterial origin of replication to allow maintenance and selection of *E. coli.* Lastly, the hygromycin B resistance gene is added to the plasmid to permit selection of transfected human cells. Into the EBV plasmid is cloned a recombinant AAV genome as illustrated by the neomycin phosphotransferase gene (NPT) driven by the murine sarcoma virus long terminal repeat (*MSV promoter*). The inverted terminal repeats of AAV (AAVITR) surround the transgene expression cassette to provide an AAV amplification unit which can subsequently be replicated and packaged in recombinant virus

transfected into human 293 cells. When Hirt extracts from individual hygromycin resistant clones were transformed into *E. coli*, 17/17 of the 293 cell lines produced bacterial colonies containing the original starting AAV/EBV chimeric plasmid. In 2/17 instances, a deleted form of the plasmid was also established in the 293 cell clones, the deletion of which most likely occurred during the initial transfection process (see below). To examine the stability of the AAV/EBV plasmids, 293 cell clones were maintained for at least 6 months in culture with hygromycin B and tested periodically for the presence and form of the chimeric plasmid. Clones that originally contained intact forms of the plasmid maintained intact forms of the plasmid during all subsequent restriction enzyme assays. In no instance were rearranged forms of the plasmid seen in lines that originally contained only intact copies. These results indicate that once the AAV/EBV plasmid was established in the 293 cells, subsequent replication of the plasmid occurred without accumulative gross rearrangements over time. We are currently analyzing plasmid copy number in these clones over extended time and determining whether periodic recloning is important to maintain high copy number.

In every case described above, the rAAV vector cloned into the EBV plasmid was defective for the *rep* functions of AAV. We have attempted to establish fully *rep*$^+$ AAV/EBV chimeric plasmids in 293 cells. Although these *rep*$^+$ plasmids are stable in *E. coli*, we have been unable to establish 293 cell lines that maintain these episomes. In the rare instances in which hygromycin resistance clones were established and EBV plasmids were reisolated, deletions in the *rep* region were observed. We have made similar observations using alternative EBV plasmid constructs in which the *rep* genes were inserted outside of the rAAV vector cassette. These observations provide support for the lethality of continuous *rep* expression in 293 cells and enforce the need for new cell lines or nonleaky regulatable promoter systems which will permit controlled expression of *rep*.

We have investigated whether 293 cell lines containing AAV/EBV chimeric plasmids can be used for the production of rAAV virus (Fig. 3). For these investigations, the 293 clones were infected with adenovirus and subsequently transfected with pΔBal, a plasmid containing the wild-type AAV functions, but lacking 121 bases of each of the AAV ITRs. Some 24 h after transfection of pΔBal, activation of rAAV replication was evident as observed by Southern blot analysis of Hirt extracts from these cells. When the Southern blots were probed with the appropriate rAAV insert, bands the size of the expected double-stranded and single-stranded monomer forms of the rAAV genome were evident. The appearance of these bands was completely dependent on the presence of the AAV/EBV chimeric plasmid and the wild-type AAV functions. When plasmids deficient for *rep* were used for complementation, or 293 lines lacking the AAV/EBV chimeric plasmids were tested, no replicating rAAV genomes were observed.

The replicating rAAV genomes were packaged into transducing virions when adenovirus and the wild-type AAV functions were supplied to 293 lines harboring

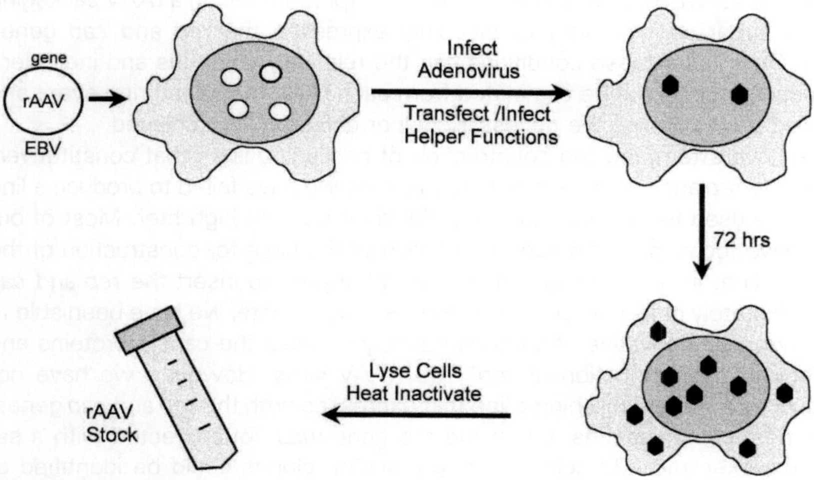

Fig. 3. Production of recombinant adeno-associated virus (rAAV) virus from 293 cells containing adeno-associated virus/Epstein-Barr virus (AAV/EBV) plasmids. 293 cell clones containing an AAV/EBV plasmid (Fig. 2) were transfected with pΔBal (LEBKOWSKI et al. 1988), a plasmid containing the *rep* and *cap* expression units of AAV but lacking the AAV inverted terminal repeats (ITRs). In addition, these cells were infected with adenovirus. The cell lysates were harvested 72 h later and tested for replicative rAAV genomes by Southern blot analysis and recombinant virus by subsequent transduction of K562 cells

AAV/EBV plasmids. Using this methodology, rAAV virus particles containing either the neomycin phosphotransferase gene or the apolipoprotein A1 gene were packaged as recombinant virions capable of infecting new cells. The transducing virus titer in crude producer cell supernatants ranged from $1 \times 10^4 - 2 \times 10^5$ transducing virions per milliliter. The rAAV titers produced by specific 293 clones were not the same and not random. In general, lines that produced higher titers continued to produce higher rAAV titers when compared to other AAV/EBV transformed 293 cell clones. The reason for this titer variability is still unknown, and we are investigating whether AAV/EBV plasmid copy number is correlative with subsequent transducing virus titer. We are also investigating whether the transfection efficiency of each of the lines influences the transducing virus titer. In all of the examples above we have used the hygromycin resistance gene to select 293 cells which contain the AAV/EBV plasmids. We have observed that hygromycin resistant 293 cells are much more difficult to transfect, even in the absence of the drug, and some of the variability in rAAV production could be a result of variable delivery or expression of the *rep* and *cap* genes. We are now looking at alternative methods to distinguish AAV/EBV plasmid transfectants and characterize clones that will be good candidate rAAV producer lines.

The above results establish that AAV/EBV plasmids can be maintained long term in human 293 cells. Moreover, upon provision of the adenovirus and wild-type AAV function to these cells, rAAV replication dominates with the accumulation of transducing virus. However, in order to take full advantage of this system, the AAV/EBV plasmid needs to be incorporated within a rAAV packaging line that either constitutively or inducibly expresses the *rep* and *cap* genes (Fig. 4). Only under these conditions can the relatively laborious and inefficient transfection procedures be eliminated from each rAAV manufacturing event and controlled production of the necessary helper cofactors be achieved.

We have attempted the construction of packaging lines that constitutively express the *rep* and *cap* gene functions. To date we have failed to produce a line that can be used to manufacture a *rep* deficient rAAV at high titer. Most of our efforts have focused on the use of 293 cells as the base for construction of the packaging line. In these investigations we attempted to insert the *rep* and *cap* genes separately or as part of the same construct. To date, we have been able to isolate two 293 derivatives that constitutively express the capsid proteins and complement the production of *rep*$^{+}$ *cap*$^{-}$ rAAV virus. However, we have not been successful in establishing a line that expresses both the *rep* and *cap* genes. Under most circumstances, when the rep gene was cotransfected with a selectable marker into 293 cells, no drug resistant clones could be identified or isolated. In a few instances, rare drug resistant clones were identified and grown to mass culture. When screened for the production of Rep, either no *rep* mRNA expression was observed or the lines failed to support the production of *rep*$^{-}$ AAV vectors. Recent evidence now suggests that 293 cells may not be the optimal line for rAAV manufacture and that *rep* expression may be better tolerated by other human cell lines. Our efforts now focus on the exploration of

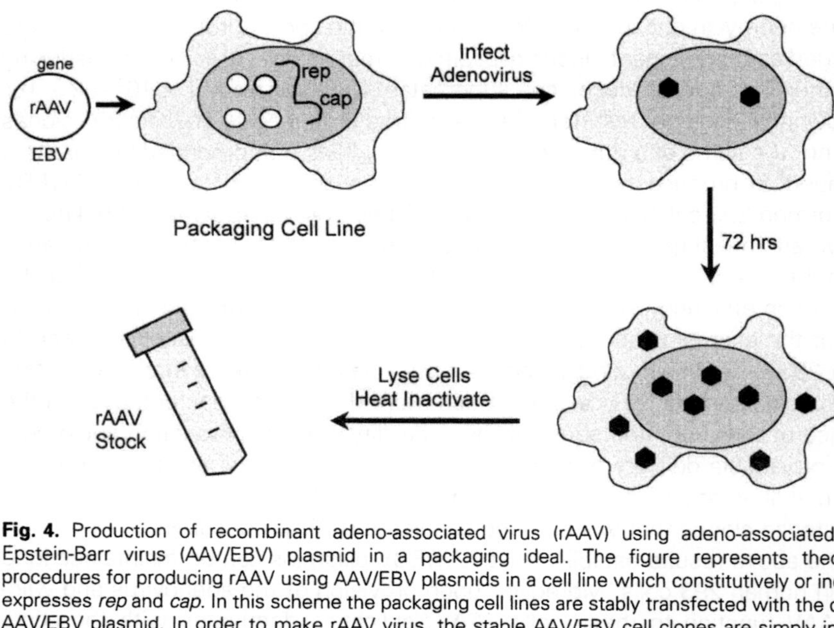

Fig. 4. Production of recombinant adeno-associated virus (rAAV) using adeno-associated virus/Epstein-Barr virus (AAV/EBV) plasmid in a packaging ideal. The figure represents theoretical procedures for producing rAAV using AAV/EBV plasmids in a cell line which constitutively or inducibly expresses *rep* and *cap*. In this scheme the packaging cell lines are stably transfected with the desired AAV/EBV plasmid. In order to make rAAV virus, the stable AAV/EBV cell clones are simply infected with adenovirus and cell lysates are collected 24 h later for subsequent purification

new cell types for the creation of packaging lines and new vector/expression systems for the delivery and expression of *rep* and *cap* in these lines.

In the meantime, several additional questions need to be answered regarding the use of rAAV vectors for gene therapy. Questions remain regarding the efficiency of recombinant virus integration and what factors influence integration of rAAV vectors. Uniform methods to titer rAAV virus must be developed and their relationship to efficient gene delivery to a targeted cell type must be established. Additional questions exist regarding the stability of transgene expression from rAAV vectors in specific tissue types. Moreover, extensive controlled studies are required to examine the storage and stability of the rAAV virions. In addition, some concerns remain regarding the safety of rAAV administration and whether rAAV production could occur in vivo upon the rare event of simultaneous infection of transduced cells with adenovirus and wild-type AAV. Finally, the efficacy of repeated administration of rAAV vectors for some therapeutic modalities must be addressed.

3 Adeno-associated Virus-Based Plasmids for Gene Transfer and Therapeutic Applications

During experiments aimed at optimizing gene transfer and expression in primary cells, we observed some unique properties of plasmids containing the ITRs of AAV. In direct comparisons with identical plasmids lacking the ITRs of AAV, transgene expression from the ITR bearing plasmids was higher and of longer duration in most cell types (PHILIP et al. 1994). Depending on the cell type and the doubling time of the cell, transgene expression from the AAV based plasmid was 50%–500% higher and 25%–300% longer than that from the standard plasmid. These observations were made in established cell lines and primary tumor and T cell samples. This effect was dependent upon the presence of two ITRs surrounding the transgenic expression cassette, although was still observable with insert sizes of 8 kb between the ITRs.

The mechanism by which the enhanced levels and duration of transgene expression are achieved is yet to be fully elucidated. We do see increased levels of transgene mRNA with plasmids containing the AAV ITRs. In addition, we have observed that plasmids containing the AAV ITRs have greater durability in transfected cells (PHILIP et al. 1994). When tested by Southern blotting analyses in time-course studies in several cell lines, the AAV plasmid as determined by Southern blotting analysis was detected for longer periods of time than were identical plasmids lacking the ITRs. After further analysis using the DNA methylation sensitive MboI and DprI restriction enzymes, no evidence for the replication of the AAV plasmid was observed in any of the cell lines tested. Moreover, extensive studies failed to show a dramatic increase in integration frequency with the AAV plasmids. One possible interpretation of these collective results is that the AAV ITRs enhance plasmid maintenance in the cells, thereby allowing greater mRNA production for longer periods of time. Further experimentation will be required to explore the mechanisms underlying these observations to date.

Many of the properties of these plasmids make them ideal vectors for a variety of gene therapy applications. We have used AAV plasmids containing the interleukin-2 (IL2) gene to transfect fresh breast, ovarian, and lung tumor cells without the need for prior ex vivo culture. Such transfection procedures are especially advantageous in immunization schemes in which culture induced alterations of the tumor cells might be detrimental. AAV plasmids carrying the IL2 gene have now been used successfully to modify tumor cells and immunize tumor bearing animals. Reductions in metastatic tumor burden have been observed in breast, ovarian, lung and prostate cancer animal models.

The human clinical use of AAV plasmids to introduce the IL2 gene into escalating doses of tumor cell vaccines is imminent. We are also exploring additional applications in which the use of transient but extended duration gene expression is advantageous, such as the modification of adoptively transferred T cells or in vivo applications for the treatment of restenosis or allograft rejection. Further explorations also focus on the discovery of methods to improve or target the integration of the AAV plasmids.

4 Conclusions

Recombinant AAV virus vectors have potential for the high efficiency transduction of numerous cell types both in vivo and ex vivo. Major efforts currently focus on improvements in high titer production of purified recombinant virus as will be required for efficient transduction and widespread safe application. Alternatively, unique properties of plasmids containing the ITRs of AAV also confer particular advantages to these vectors for many transient gene expression applications. The use of these plasmids for single or repetitive dose transient expression reduces the complexity of vector production and decreases the antigenic potential of the gene transfer agent compared to viral particles. The plasmid- and virus-based vectors have distinct advantages and targeted application in a variety of clinical scenarios. Upcoming clinical programs using AAV viral and plasmid vectors will be especially informative and provide information for future research on these gene transfer systems.

References

Berns KI, Bohensky RA (1987) Adeno-associated viruses: an update. Adv Vir Res 32: 243–306

Flotte TR, Barraza-Ortiz X, Solow R, Afione SA, Carter BJ, Guggino WB (1995) An improved system for packaging recombinant adeno-associated virus vectors capable of in vivo transduction. Gene Ther 2(1): 29–37

Holscher C, Hoier M, Kleinschmidt JA, Zentgral H, Burkle A, Heilbronn R (1994) Cell lines inducibly expressing the adeno-associated virus (AAV) rep genes: requirements for productive replication of rep-negative AAV mutants. J Virol 68(11): 7169–7177

Lebkowski JS, McNally M, Okarma TB, Lerch LB (1988) Adeno-associated virus: a vector system for efficient introduction and integration of DNA into a variety of mammalian cell types. Mol Cell Biol 8: 3988–3996

Mamounas M, Leavitt M, Yu M, Wong-Staal F (1995) Increased titer of recombinant AAV vectors by gene transfer with adenovirus coupled to DNA-polylysine complexes. Gene Ther 2: 429–432

Musycka N (1992) Use of adeno-associated virus as a general transduction vector for mammalian cells. In: Muzyczka N (ed) Viral expression vectors. Springer, Berlin Heidelberg New York, pp 97–129 (Current topics in microbiology and immunology, vol 158)

Philip R, Brunette E, Kilinski L, Murugesh D, McNally MA, Ucar K, Rosenblatt J, Okarma TB, Lebkowski JS (1994) Efficient and sustained gene expression in primary T lymphocytes and primary and cultured tumor cells mediated by adeno-associated virus plasmid DNA complexed to cationic liposomes. Mol Cell Biol 14(4): 2411–2418

Rolling F, Samulski RJ (1995) AAV as a viral vector for human gene therapy. Generation of recombinant virus. Mol Biotechnol 3(1): 9–16

Sugden B, Marsh K, Yates J (1985) A vector that replicates as a plasmid and can be efficiently selected in B-lymphoblasts transformed by Epstein-Barr virus. Mol Cell Biol 5: 410–413

Yang Q, Chen F, Trempe JB (1994) Characterization of cell lines that inducibly express the adeno-associated virus rep proteins. J Virol 68(8): 4847–4856

Yates JL, Warren N, Sugden B (1985) Stable replication of plasmids derived from Epstein-Barr virus in various mammalian cells. Nature 313: 812–815

Adeno-associated Virus Vectors for Gene Therapy of the Hematopoietic System

S. CHATTERJEE[1] and K.K. WONG JR.[2]

1 Introduction

Over the past 30 years, bone marrow transplantation has come to be accepted as standard curative therapy for many pathophysiologic conditions (FORMAN et al. 1994) including malignancies (BRENNER et al. 1993) and metabolic diseases (BLAESE 1993). Thus, the reconstitution of the hematopoietic system with normal allogeneic cells capable of providing long term engraftment with disease-free, physiologically functional cells has been used to treat various inherited diseases including severe combined immunodeficiency disease (SCID), lysosomal storage disorders and various hemoglobinopathies. However, difficulties in acquisition of histocompatible marrow cells for transplantation and onset of graft vs host disease has led to the continued search for better therapeutic strategies. Meanwhile, the delineation of the genetic bases of many inherited diseases and the subsequent isolation of appropriate wild-type genes promoted the feasibility of disease correction at the DNA level. To this end, much effort has focussed

[1]Division of Pediatrics, City of Hope National Medical Center, 1500 E. Duarte Road, Duarte, CA 91010, USA
[2]Department of Hematology and Bone Marrow Transplantation, City of Hope National Medical Center, 1500 E. Duarte Road, Duarte, CA 91010, USA

upon the development of both the identification of genetic strategies as well as efficient methods of delivery of therapeutic genes to appropriate target cells.

Cells of the hematopoietic system readily lend themselves to ex vivo gene therapy. The bone marrow is an accessible source of lineage-committed progenitor cells and, importantly, pluripotent hematopoietic stem cells theoretically capable of both self-renewal and multilineage reconstitution of the entire hematopoietic system with gene-modified cells. Thus, the ability to provide long term correction of inherited diseases through the genetic modification of hematopoietic stem cells is attractive. However, the precise identification and purification of pluripotent stem cells have remained elusive. Bone marrow, cytokine-mobilized peripheral blood and umbilical cord blood have all been identified as sources of hematopoietic stem and progenitor cells. Recent studies have demonstrated that cell populations exhibiting high levels of cell surface expression of the CD34 differentiation antigen and low expression of thy 1 antigen, and lacking lineage-specific differentiation markers contain primitive cells capable of giving rise to cells of lymphoid, erythroid and myeloid lineages in SCID mice repopulated with human cells (Baum et al. 1992).

Primitive hematopoietic cells capable of multilineage differentiation have been shown to be largely quiescent with only a few cells being in active cycle at any given time (Ogawa 1993). Stem cell division is generally associated with both differentiation along lineage pathways as well as self-renewal. Thus for permanent gene therapy of the hematopoietic system, gene transfer into self-renewing pluripotential stem cells is highly desirable to ensure the continued presence of the transgene in terminally differentiated progeny while maintaining a pool of genetically modified progenitor cells. For effective stem cell gene therapy it is also important to achieve permanent transgene integration into the stem cell genome to ensure transmission to progeny. Of the currently available gene transfer systems only two offer the possibility of stable transgene integration. These include viral vectors based upon retroviruses and adeno-associated viruses (AAVs). Other gene transfer methods, including adenovirus vectors and liposome-based DNA delivery methods, do not result in efficient transgene integration. Thus the major utility of these systems may lie in gene therapy strategies requiring transient transgene expression. This chapter focusses upon integrating vectors for the ultimate purpose of stem cell gene therapy.

2 Retroviral Vectors

Retroviral vectors based upon Moloney murine leukemia virus (MLV) have been studied extensively for their capacity to transfer genes into hematopoietic progenitor cells (Miller et al. 1993). Vectors are constructed by replacement of viral genes with heterologous open reading frames downstream of the retroviral long terminal repeats (LTRs) (Fig. 1). Retention of the packaging signal assures effi-

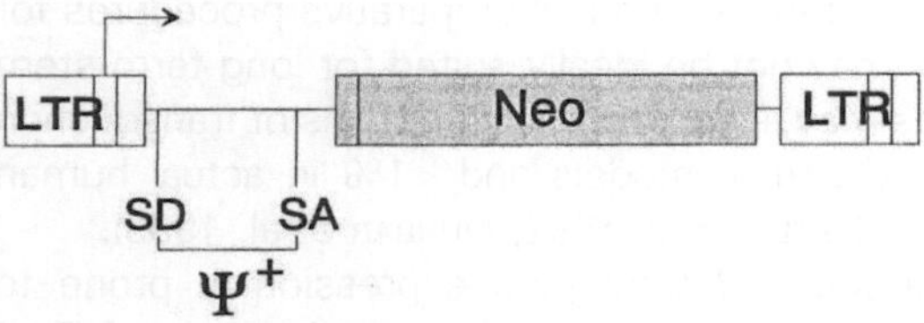

Fig. 1. Map of a prototypic retroviral vector with transgene expression under long terminal repeat (LTR) control

cient encapsidation of genomic viral RNA into core particles. The vectors are packaged in producer cell lines permanently transformed with retroviral *gag, pol* and *env* genes as well as vector sequences (MILLER 1990). Derivation of producer cells requires the presence of a selectable marker in the vector to aid in clonal selection of producer lines. Infectious retroviruses are produced by budding from the plasma membrane into the tissue culture supernatant. Gene transductions are performed by addition of vector-containing supernatant to target cells or by coculture of target cells with producer cells. Transgene expression is under the control of either the retroviral LTR or an internal promoter such as the SV40 or cytomegalovirus early promoter. Some retroviral constructs with tRNA promoters driving expression of short biologically active RNA molecules have also been reported. These have been placed within the LTRs to avoid promoter interference and increase expression (SULLENGER et al. 1990).

Gene therapy strategies utilizing retroviral vectors are currently being developed for the hematopoietic system for a variety of diseases, including acquired immunodeficiency disease (AIDS) (YU et al. 1994), Gaucher's disease (BEUTLER 1993), adenosine deaminase deficiency (BLAESE 1993), as well as for the genetic marking of transplanted cells during cancer therapy (BRENNER et al. 1993). Retroviral vectors are well defined and currently represent the most commonly utilized viral vector system in clinical trials. However retroviral vectors have a requirement for active target cell division (MILLER et al. 1990) for efficient transduction. Mitosis-induced nuclear membrane breakdown is necessary for the entry of the pre-integration complex into the nucleus (ROE et al. 1993). This inability to transduce nondividing cells is a major impediment for gene transfer into primary cells as reflected in reports of low retroviral gene transfer frequencies into hematopoietic stem cells in vivo in large animal models. Thus, while gene transfer into cytokine or chemotherapy-stimulated progenitor cells has been accomplished with retroviruses, gene transfer into unmanipulated primitive stem cells has remained elusive. Cytokine stimulation of primitive stem cells, either by the addition of exogenous cytokines (LUSKEY et al. 1992) or by growth on stromal cell layers (NOLTA et al. 1992), has now become standard procedure for retroviral transduction. However, cytokines that stimulate stem cells to proliferate without concomitant differentiation have yet to be identified. Cytokine combinations defined to date all result in the loss of self-renewal capacity and in lineage commitment (OGAWA 1993) of primitive hematopoietic stem cells. Thus cytokine-facilitated gene transfer and cell preparative procedures for retrovirus-mediated gene therapy may not be ideally suited for long term stem cell transduction. These limitations may account for observations of transduction frequencies of <5% in canine and simian models and <1% in actual human clinical trials. (KANTOFF et al. 1987; CARTER et al. 1992; DUNBAR et al. 1995).

Additionally, retroviral vector-encoded transgene expression is prone to down-regulation, possibly due to mechanisms including methylation of CpG sequences in the MLV LTRs (CHALLITA and KOHN 1994), which act as promoter elements directing transgene expression or by heterochromatin formation. Thus in addition to low transduction frequencies, sustained gene expression of retroviral -encoded transgenes has been problematic, and the continued search for other novel vectors for use in stem cell gene therapy is imperative.

3 Adeno-associated Virus Vector Biology

Adeno-associated virus is a single-stranded, replication-defective DNA virus with a 4.7 kb genome with palindromic inverted terminal repeats (ITRs) (BERNS and BOHENZKY 1987; MUZYCZKA 1992). Coinfection with a helper virus, such as adenovirus, is required for productive infection (JANIK et al. 1981; BULLER et al. 1981). Molecular clones of the AAV genome are infectious following transfection into helper virus infected cells (SAMULSKI et al. 1982; LAUGHLIN et al. 1983), facilitating genetic manipulations. In the absence of helper virus, AAV integrates in a stable fashion via the ITRs into human chromosomal DNA (BERNS et al. 1975). Wild-type AAV has recently been shown to integrate in a site-specific manner into q-ter 13.2–13.4 of the human chromosome 19 (KOTIN et al. 1990; SAMULSKI et al. 1991). While wild-type-free AAV vectors have not been shown to integrate into 19 q-ter 13.2–13.4, the possibility of directing nonrandom integration of AAV-based vectors is attractive since it would minimize the risk of insertional mutagenesis and the variability of inserted transgene expression. AAV vectors have high transduction frequencies (McLAUGHLIN et al. 1988) in cells of diverse lineages including hematopoietic cells, attractive targets for human gene therapy. Genes inserted into AAV vectors may be precisely designed to direct the transcription of discrete mRNAs. Although AAV ITRs have been shown to direct gene expression (FLOTTE et al. 1993) in the absence of internal promoters, no ITR-directed transcription is detected when internal promoters are present (BRAR et al. 1994). Thus, promoter interference does not occur with the base vector. Additionally, AAV vectors often integrate in tandem in multiple copies (McLAUGHLIN et al. 1988), thereby enhancing transgene expression. Finally, latent wild-type AAV infections have been stably maintained in tissue culture for greater than 100 serial passages in the *absence* of selective pressure, indicating the stability of

AAV genomic integration (BERNS et al. 1975). Thus, AAV vectors are particularly well suited for transgene delivery.

Recombinant AAV virions can efficiently package inserts up to approximately 5 kb in length. The packaged DNA is single-stranded with both strands being packaged in separate virions. Only the 5'- and 3'- ends of the viral genome which include the 145 base long palindromic ITRs are necessary for efficient packaging (SAMULSKI et al. 1989). The remainder of the recombinant AAV vector genome may carry heterologous sequences in any transcriptional units desired. We have designed vectors based that contain bases 1–189 and 4499–4680 derived from wild-type AAV-2 and include both ITRs and the endogenous AAV polyadenylation and RNA termination signals (WONG et al. 1991; CHATTERJEE et al. 1992). Drug resistance genes, such as the neomycin phosphotransferase or the hygromycin B phosphotransferase genes, under the control of separate promoters and carrying its own polyadenylation signal, may be included to allow positive selection of transduced cells.

Vectors based upon AAV have been constructed by the replacement of endogenous viral genes with genes of interest under transcriptional control of a variety of promoters and utilizing separate RNA termination and polyadenylation signals (Fig. 2). Promoters shown to be active in AAV vectors include the Rous sarcoma virus (RSV) LTR, the SV40 early promoter, the cytomegalovirus immediate early promoter and the murine leukemia virus LTR (LEBKOWSKI et al. 1988; WONG et al. 1991; CHATTERJEE et al. 1992; MUZYCZKA 1992; CHATTERJEE and WONG 1993; LU et al. 1994a,b; BRAR et al. 1994; RUSSELL et al. 1994). In one

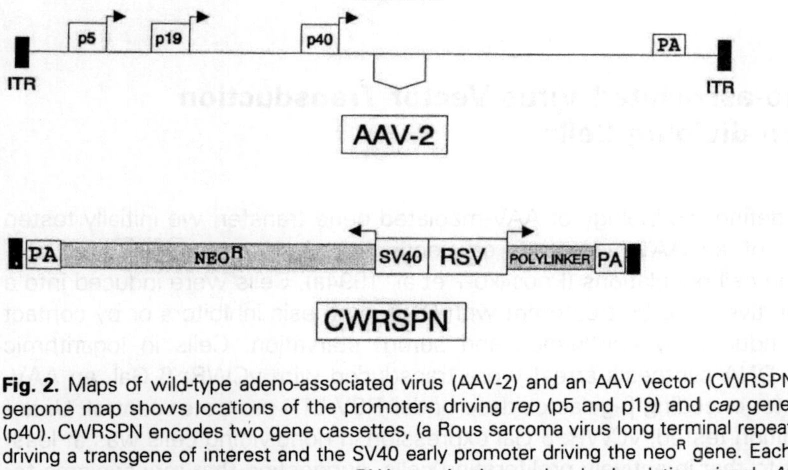

Fig. 2. Maps of wild-type adeno-associated virus (AAV-2) and an AAV vector (CWRSPN). The AAV genome map shows locations of the promoters driving *rep* (p5 and p19) and *cap* gene expression (p40). CWRSPN encodes two gene cassettes, (a Rous sarcoma virus long terminal repeat) (RSV LTR) driving a transgene of interest and the SV40 early promoter driving the neo^R gene. Each expression unit in the vector utilizes a separate RNA termination and polyadenylation signal. *Arrows* show direction of transcription. The kilobase scale indicates approximate sizes of genes

instance, the ITR has been utilized to drive the cystic fibrosis transmembrane conductance regulatory gene (FLOTTE et al. 1993). Multiple genes, each under separate transcriptional control are functional in AAV vectors. RNA polymerase II and III-dependent promoters are active in single and multigene constructs (CHATTERJEE and WONG 1993; BERTRAND et al. 1996). A major limitation of AAV vector constructs is in the size of the vector genome. AAV vectors have been shown to efficiently encapsidate up to 5 kb of single-stranded DNA.

Most AAV vector packaging procedures reported to date involve transfection of the vector plasmid along with another plasmid providing AAV *rep* (DNA replication) and *cap* (virion proteins) genes into helper virus-infected cells. Cell-associated recombinant vectors are harvested 48–72 h after transfection. The cells are lysed and processed to release virions and after heat inactivation of helper virus, the crude cell lysates serve as a source of transducing vector (CHATTERJEE and WONG 1993). Like wild-type AAV, AAV vectors appear to be resistant to a variety of treatments including certain lipid solvents and heating to 56 °C for up to 10 h. These properties may be used to further purify virions away from helper viruses. AAV vectors may also be further purified on isopycnic cesium chloride gradients in which wild-type AAV bands at a density of 1.4. The density of vectors is dependent upon the size of the genome.

AAV vectors have a wide host range and tissue tropism and, like the wild-type virus, integrate into host chromosomal DNA. We have recently utilized viral vectors based upon AAV to efficiently confer intracellular resistance to human immunodeficiency virus (HIV-1) and herpes simplex virus (HSV-1) (WONG et al. 1991; CHATTERJEE et al. 1992). To further develop AAV vector systems for gene therapy of HIV infection and oncogenesis, we are investigating AAV-mediated gene transfer into human hematopoietic progenitor cells.

4 Adeno-associated Virus Vector Transduction of Non-dividing Cells

To better define the biology of AAV-mediated gene transfer, we initially tested the ability of an AAV vector to efficiently introduce transgenes into non-proliferating cell populations (PODSAKOFF et al. 1994a). Cells were induced into a nonproliferative state by treatment with DNA synthesis inhibitors or by contact inhibition induced by confluence and serum starvation. Cells in logarithmic growth or DNA synthesis arrest were transduced with vCWR: β Gal, an AAV-based vector encoding β-galactosidase under RSV LTR promoter control. Under each condition tested, vCWR: β Gal expression in nondividing cells was at least equivalent to that in actively proliferating cells, suggesting that mechanisms for virus attachment, nuclear transport, virion uncoating and perhaps some limited second strand synthesis of AAV vectors were present in nondividing cells. Southern hybridization analysis of vector sequences from cells transduced while in DNA synthetic arrest and expanded after release of block confirmed *ultimate* integration of the vector genome into cellular chromosomal DNA. It is currently unclear as to exactly when AAV vectors integrate into chromosomal DNA. However these results suggest that, unlike retroviral vector genomes, AAV genomes are capable of surviving in the cell until a milieu permissive for integration is achieved.

Unlike the findings of ALEXANDER et al. (1994), in which the induction of DNA damage was required for efficient AAV transduction of nondividing human diploid fibroblasts, our results and those of others indicate that transduction efficiencies ranging from 50%–100% are observed in a reproducible fashion in unmanipulated cells (CHATTERJEE et al. 1992). Furthermore, our results were recently confirmed by FLOTTE et al. (1994), who also showed sustained gene expression in nondividing respiratory epithelial cells, and KAPLITT et al. (1994), who reported stable AAV-mediated gene transfer and long term transgene expression in postmitotic neurons of adult rats. One laboratory initially reported conflicting results (RUSSELL et al. 1994); but subsequent papers from the same group showed that cells arrested in mitosis with inhibitors such as methotrexate and nocodazole were transducible with AAV (ALEXANDER et al. 1994). Thus these findings suggest the utility of AAV-based vectors for gene transfer into quiescent cell populations.

5 Adeno-associated Virus Transduction of CD34⁺ Hematopoietic Progenitor Cells

To evaluate the utility of AAV vectors for use in gene therapy of the hematopoietic system, we and others have tested AAV-mediated gene transfer into human CD34$^+$ progenitor cells. Without further fractionation, this hematopoietic cell population consists primarily of lineage committed progenitor cells. Thus lymphoid, myeloid and erythroid progenitor cells all express the CD34 antigen on their surface. However approximately 0.1%–1.0% of total CD34$^+$ cells are thought to represent pluripotent progenitors capable of self-renewal and differentiation along any of the hematopoietic lineages. Therefore gene transfer into the total population of CD34$^+$ cells may only test transduction of committed progenitor cells with limited life spans and may not necessarily reflect long term stem cell gene transfer. Nevertheless, CD34$^+$ cells are an easily accessible source of progenitor cells for initial evaluation of AAV vectors for hematopoietic gene therapy.

Our group has used vectors encoding either reporter genes or an antisense RNA to the HIV-1 LTR under control of various viral or cellular promoters to evaluate transduction of CD34$^+$ cells. CD34$^+$ cells are obtained by either immunomagnetic bead separation or by panning on antibody-coated flasks or columns, transduced with recombinant AAV vectors and placed in culture in media containing cytokines. The presence of cytokines is required for maintenance of

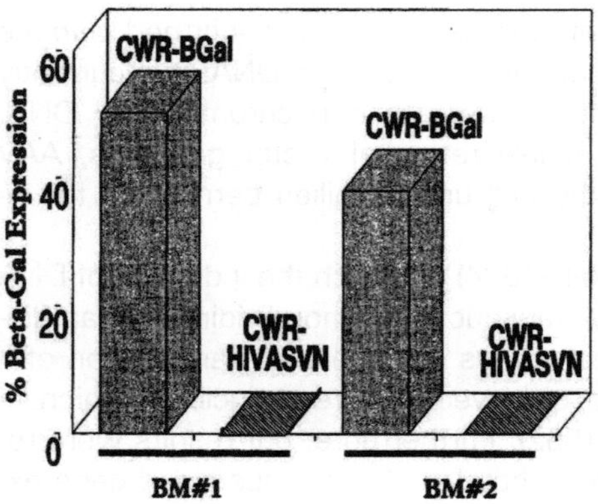

Fig. 3. Transduction of CD34 cells with vCWR-BGal, an adeno-associated virus vector encoding the β-galactosidase gene under RSV LTR control. *Bars* show β-galactosidase expression in CD34 cells 7 days posttransduction as assessed by histochemical stain. CWR-HIVASVN-transduced cells serve as negative controls. BM1 and BM2 represent cells obtained from two different marrow samples

progenitor cell viability; however, the cytokines identified to date are known to induce differentiation and cell proliferation (OGAWA 1993) and do not support the expansion of stem cells. Figure 3 shows an example AAV-mediated gene transfer into marrow-derived CD34$^+$ cells assessed 7 days after transduction. vCWR-BetaGal was used to transduce CD34$^+$ cells from two individuals at a multiplicity of 0.5. Approximately 40%–50% of the cells exhibited transgene expression, while control transduced (CWRHIVASVN) or untransduced cells showed no beta-galactosidase activity. This suggested the feasibility of continuing the evaluation of AAV-mediated gene transfer into hematopoietic progenitor cells. Subsequent experiments with different AAV vectors showed that AAV vector transduction at multiplicity of ~0.5–1 resulted in transgene expression in 40%–90% of CD34$^+$ cells in short term suspension cultures. These results indicated that mechanisms for entry, nuclear translocation, second strand DNA synthesis and gene expression from AAV vectors were present in CD34$^+$ cells. Evaluation of transduction of myeloid progenitor cells capable of forming colonies in vitro (CFU-C) from transduced CD34$^+$ cells demonstrated specific transgene expression with an efficiency of 50%–90%. Integration analysis of transduced CD34$^+$ cell DNA revealed the presence of integrated vector sequences in chromosomal DNA, indicating that AAV transduction resulted in stable gene transfer (CHATTERJEE et al. 1995; FISHER-ADAMS et al. 1996).

Table 1 summarizes results obtained to date with AAV transduction of hematopoietic progenitor cells. ZHOU et al. (1994) have also demonstrated efficient AAV transduction of myeloid progenitor cells derived from cord blood. Wild-type AAV infection and transduction of nonhuman primate CD34$^+$ cells was demonstrated by GOODMAN et al. (1994). WALSH et al. (1994) demonstrated AAV transduction of the gene encoding the Fanconi anemia complementation group C into peripheral blood CD34 cells by clonogenic assays and RNA PCR analysis. MILLER et al. (1994) described AAV transduction of the γ-globin gene into human erythroid colony forming cells. In contrast to the findings of HALBERT et al. (1995), our results and those of others suggest that AAV vectors transduce primary human cells at high efficiencies.

Table 1. Transduction of hematopoietic cells with adeno-associated virus vectors

Target cell	Transduction efficiency	Transgene	Reference
Murine marrow mononuclear cells	50%–80%	Neo^{R1}	ZHOU et al. 1993
Human umbilical cord blood $CD34^+$ cells	33%–75%	Neo^{R1}	ZHOU et al. 1994
Human peripheral blood $CD34^+$ cells	Up to 60%	$FACC^2$	WALSH et al. 1994
Human peripheral blood $CD34^+$ cells	20%–30%	Human γ-globin	MILLER et al. 1994
Human marrow-derived $CD34^+$ cells	>70%	$PLAP^3$, Neo^R Antisense RNA	FISHER-ADAMS et al. 1996

[1]Neomycin phosphotransferase.
[2]Fanconi anemia, complementation group C gene.
[3]Human thermostable placental alkaline phosphatase.

Optimal retroviral vector-mediated gene transfer into CD34 cells requires prior cytokine stimulation for 48–72 h. Thus cytokine-facilitated retroviral transduction results in gene transfer into lineage-committed cells. Thus for the development of stem cell gene therapy, it was important to evaluate the capacity of AAV vectors to mediated gene transfer into $CD34^+$ cells prior to cytokine stimulation. Both our group (FISHER-ADAMS et al. 1996) and ZHOU et al. (1994) have found comparable levels of transgene expression whether the cells are transduced prior to or after cytokine stimulation. Interestingly, the proportion of dividing cells in the $CD34^+$ population is known to be small. While these results do not prove transduction of quiescent cells, they are nevertheless consistent with it.

6 Adeno-associated Virus Transduction of Primitive Hematopoietic Precursor Cells

The lack of assay systems of directly test self-renewing, multilineage, pluripotential human stem cells in vitro has impeded direct attempts to analyze gene transfer into this important cell population. However, long term marrow cultures in which a microenvironment capable of supporting hematopoiesis for up to 8 weeks, have been described (DEXTER et al. 1977). Long term cultures exploit the ability of marrow-derived adherent cells to support the viability of hematopoietic progenitor cells. The cultures are initiated by plating CD34 cells on an irradiated, heterologous, marrow-derived, stromal cell layer. Primitive progenitor cells grow in association with stromal cells, forming "cobblestone areas." The addition of cytokines is not required for the maintenance of primitive progenitor cells in LTC since stromal cells provide both secreted and contact-dependent factors necessary for optimal progenitor cell viability. In this culture system, the committed progenitor cells die out in the first 3 weeks of culture while clonogenic cells detectable at later time points represent the progeny of more primitive cells (SUTHERLAND et al. 1990). These cells, known as long term culture initiating cells (LTC-ICs), show good correlation with long term marrow repopulating activity and represent the most primitive human hematopoietic progenitor cell currently assayable in vitro (SUTHERLAND et al. 1991). Clonogenic cells from long term cultures have been shown to contain precursors of erythroid and myeloid lineages. Analysis of gene transfer into LTC and LTC-ICs indicated stable transduction, with vector sequences being detectable up to 7–8 weeks after transduction at frequencies of 0.2–1 copy per cell in the absence of selective pressure (LU et al. 1994a,b). Analysis of transgene expression both at the protein and RNA levels suggested that significant proportions of primitive human hematopoietic progenitor cells were transduced by AAV even in the absence of selective pressure. These results indicate that AAV transduction of primitive human myeloid progenitor cells is efficient.

However, even LTC and LTC-IC systems do not assay for precursors of lymphoid cells, the earliest lineage thought to differentiate from stem cells. Thus analysis of true pluripotent stem cells requires in vivo assays of transduction. Long term multilineage hematopoietic reconstitution of lethally irradiated animals ultimately provides the best test of stem cell transduction. We therefore tested the ability of AAV-transduced murine marrow cells to engraft lethally irradiated syngeneic mice. Early results demonstrated that mice transplanted with AAV-transduced marrow cells showed evidence of long term, multilineage reconstitution (PODSAKOFF et al. 1994b). Transgene-expression cells was detected in peripheral blood and hematopoietic organs at >6 months post-transplantation, with no detectable hematologic toxicity due to AAV transduction. Retroviral vector transduction of murine hematopoietic progenitor cells is efficient; however, in large animal models gene transfer frequencies are very low. In human clinical trials, retroviral transduction is reported to occur at a frequency of 1:10 000 (DUNBAR et al. 1995). Therefore the continued investigation of AAV-mediated gene transfer into primary hematopoietic progenitor cells is imperative and may provide the basis for novel modes of ex vivo gene therapy of the hematopoietic system.

Conclusions

Adeno-associated virus vectors are rapidly emerging as promising vectors for use in human gene therapy by virtue of their high transduction frequencies in primary cells and ability to transduce nondividing cells and primitive hematopoietic progenitor cells. Transcription of transgenes from AAV vectors is efficient and sustained. These properties render AAV vectors as good candidates for therapeutic gene transfer for the treatment of hereditary diseases (MILLER et al. 1994; WALSH et al. 1994), antiviral approaches to diseases such as AIDS (CHAT-

TERJEE et al. 1992), anti-oncogene strategies (Lu et al. 1994a,b) and marking studies to detect minimal residual disease in stem cell transplants. As applications for human gene therapy expand, the particular biological properties of each vector system may be exploited for use in specific situations. The high efficiency of AAV-mediated gene transfer into primary human cells reported thus far makes it compelling to further define the biology of this system.

References

Alexander IE, Russell DW, Miller AD (1994) DNA-damaging agents greatly increase the transduction of nondividing cells by adeno-associated virus vectors. J Virol 68: 8282–8287

Baum CM, Weissman IL, Tsukamoto AS, Buckle AM, Peault B (1992) Isolation of a candidate human hematopoietic stem-cell population. Proc Natl Acad Sci USA 89: 2804–2808

Berns K, Pinkerton T, Thomas GF, Hoggan MN (1975) Detection of adeno-associated virus (AAV)-specific nucleotide sequences in DNA isolated from latently infected Detroit 6 cells. Virology 68: 556–560

Berns KI, Bohenzky RA (1987) Adeno-associated viruses: an update. Adv Virus Res 32: 243–306

Bertrand E, Carbonelle E, Lee NS, Zhou C, Kohn D, Good P, Engelke D, Chatterjee S, Rossi J (1996) Development of viral vectors for expression of ribozyme and antisense RNAs in mammalian cells. Nucleic Acids Res (submitted)

Beutler E (1993) Gaucher disease as a paradigm of current issues regarding single gene mutations of humans. Proc Natl Acad Sci USA 90: 5384–5390

Blaese RM (1993) Development of gene therapy for immunodeficiency: adenosine deaminase deficiency. Pediatr Res 33: S49

Brar D, Wong KK Jr, Permana P, Chatterjee S (1994) Promoter interactions in adeno-associated virus vectors encoding multiple gene cassettes: potential use in anti-oncogene vector design. Cancer Gene Ther 1: 321

Brenner MK, Rill DR, Holladay MS, Heslop HE, Moen RC, Buschle M, Krance RA, Santana VM, Anderson WF, Ihle JN (1993) Gene marking to determine whether autologous marrow infusion restores long-term haemopoiesis in cancer patients. Lancet 342: 1134–1137

Buller RML, Janik J, Sebring ED, Rose JA (1981) Herpes simplex virus types 1 and 2 completely help adenovirus-associated virus replication. J Virol 40: 241–247

Carter RB, Abrams-Ogg AC, Dick JE, Kruth SA, Valli VE, Kamel-Reid S, Dube ID (1992) Autologous transplantation of canine long-term marrow culture cells genetically marked by retroviral vectors. Blood 79: 356

Challita PM, Kohn DB (1994) Lack of expression from a retroviral vector after transduction of murine hematopoietic stem cells is associated with methylation in vivo. Proc Natl Acad Sci USA 91: 2567–2571

Chatterjee S, Wong KK (1993) Adeno-associated virus vectors for the delivery of antisense RNA Methods (a companion to Methods in Enzymology) 5: 51–59

Chatterjee S, Johnson PR, Wong KK (1992) Dual target inhibition of HIV-1 in vitro with an adeno-associated virus-based antisense vector. Science 258: 1485–1488

Chatterjee S, Wong KK, Lu D, Permana P, Podsakoff GM (1995) Novel approaches for efficient gene transfer into hematopoietic progenitor cells: the use of adeno-associated virus vectors. Bone Marrow Transplant 15: 5309–5313

Dexter TM, Allen TD, Lajtha LG (1977) Conditions controlling the proliferation of haemopoietic stem cells in vitro. J Cell Physiol 91: 335–344

Dunbar CE, Cottler-Fox M, O'Shaughnessy JA, Doren S, Carter C, Berenson R, Brown S, Moen RC, Greenblatt J, Stewart FM, Leitman SF, Wilson WH, Cowan K, Young NS, Nienhuis AW (1995) Retrovirally marked CD34-enriched peripheral blood and bone marrow cells contribute to long-term engraftment after autologous transplantation. Blood 85: 3048–3057

Fisher-Adams G, Wong KK, Podsakoff GM, Forman S, Chatterjee S (1996) Integration of adeno-associated virus vector genomes in human CD34 cells following transduction. Blood 88 (in press)

Flotte TR, Afione SA, Conrad C, McGrath SA, Solow R, Oka H, Zeitlin PL, Guggino WB, Carter BJ (1993) Stable in vivo expression of the cystic fibrosis transmembrane conductance regulator with an adeno-associated virus vector. Proc Natl Acad Sci USA 90: 10613

Flotte TR, Afione SA, Zeitlin PL (1994) Adeno-associated virus vector gene expression occurs in nondividing cells in the absence of vector DNA integration. Am J Respir Cell Mol Biol 11: 517–521

Forman SJ, Blume KG, Thomas ED (eds) (1994) Bone marrow transplantation. Blackwell Scientific, Cambridge

Goodman S, Xiao X, Donahue RE, Moulton A, Miller J, Walsh C, Young NS, Samulski RJ, Nienhuis AW (1994) Recombinant adeno-associated virus-mediated gene transfer into hematopoietic progenitor cells. Blood 84: 1492–1500

Halbert CL, Alexander IE, Wolgamot GM, Miller AD (1995) Adeno-associated virus vectors transduce primary cells much less efficiently than immortalized cells. J Virol 69: 1473–1479

Janik JE, Huston MM, Rose JA (1981) Locations of adenovirus genes required for the replication of adeno-associated virus. Proc Natl Acad Sci USA 78: 1925–1930

Kantoff PW, Gillio AP, McKachlin JR, Bordignon C, Eglitis MA, Kernan NA, Moen RC, Kohn DB, Yu SF, Karson E, Karlsson S, Zwiebel JA, Gilboa E, Blaese RM, Nienhuis A, O'Reilly RJ, Anderson WF (1987) Expression of human adenosine deaminase in nonhuman primates after retrovirus-mediated gene transfer. J Exp Med 166: 219–234

Kaplitt MG, Leone P, Samulski RJ, Xiao X, Pfaff DW, O'Malley KL, During MJ (1994) Long-term gene expression and phenotypic correction using adeno-associated virus vectors in the mammalian brain. Nature Genet 8: 148–153

Kotin RM, Siniscalco M, Samulski RJ, Zhu XD, Hunter L, Laughlin CA, Mclaughlin S, Muzyczka N, Rocchi M, Berns KI (1990) Site-specific integration by adeno-associated virus. Proc Natl Acad Sci USA 87: 2211–2215

Laughlin CA, Tratschin J-D, Coon H, Carter BJ (1983) Cloning of infectious adeno-associated virus genomes in bacterial plasmids. Gene 23: 65–73

Lebkowski JS, McNally MM, Okarma TB, Lerch LB (1988) Adeno-associated virus: a vector system for efficient introduction and integration of DNA into a variety of mammalian cell types. Mol Cell Biol 8: 3988–3996

Lu D, Chatterjee S, Brar D, Wong KK Jr (1994a) High efficiency in vitro cleavage of transcripts arising from the major transforming genes of human papillomavirus type 16 mediated by ribozymes transcribed from an adeno-associated virus-based vector. Cancer Gene Therapy. 1: 267–277

Lu D, Wong KK Jr, Podsakoff G, Chatterjee S (1994b) Stable and efficient adeno-associated virus-mediated gene transfer into primitive human marrow-derived hematopoietic cells in long term culture. Blood 84 [Suppl 1]: 360a

Luskey BD, Rosenblatt M, Zsebo K, Williams DA (1992) Stem cell factor, interleukin-3, and interleukin-6 promote retroviral-mediated gene transfer into murine hematopoietic stem cells. Blood 80: 396–402

McLaughlin SK, Collis P, Hermonat PL, Muzyczka N (1988) Adeno-associated virus general transduction vectors: analysis of proviral structures. J Virol 62: 1963–1973

Miller AD (1990) Retrovirus packaging cells. Hum Gene Ther 1: 5–14

Miller DG, Adam MA, Miller AD (1990) Gene transfer by retrovirus vectors occurs only in cells that are actively replicating at the time of infection. Mol Cell Biol 10: 4239–4242

Miller AD, Miller DG, Garcia JV, Lynch CM (1993) Use of retroviral vectors for gene transfer and expression. Methods Enzymol 217: 581–599

Miller JL, Donahue RE, Sellers SE, Samulski RJ, Young NS, Nienhuis AW (1994) Recombinant adeno-associated virus (rAAV)-mediated expression of a human gamma-globin gene in human progenitor-derived erythroid cells. Proc Natl Acad Sci USA 91: 10183–10187

Muzyczka N (1992) Use of AAV as a general transduction vector for mammalian cells. In: Muzyczka N (ed) Viral expression vectors. Springer, Berlin Heidelberg New York, pp 97–129 (Current topics in microbiology and immunology, vol 158)

Nolta JA, Crooks GM, Overell RW, Williams DE, Kohn DB (1992) Retroviral vector-mediated gene transfer into primitive human hematopoietic progenitor cells: effects of mast cell growth factor (MGF) combined with other cytokines. Exp Hematol 20: 1065–1071

Ogawa M (1993) Differentiation and proliferation of hematopoietic stem cells. Blood 81: 2844–2853

Podsakoff G, Wong KK Jr, Chatterjee S (1994a) Stable and efficient gene transfer into non-dividing cells by adeno-associated virus (AAV)-based vectors. J Virol 68: 656–5666

Podsakoff G, Shaughnessy EA, Lu D, Wong KK Jr, Chatterjee S (1994b) Long term in vivo reconstitution with murine marrow cells transduced with an adeno-associated virus vector. Blood 84 [Suppl] 256a

Roe TY, Reynolds TC, Yu G, Brown PO (1993) Integration of murine leukemia virus DNA depends on mitosis. EMBO J 12: 2099–2108

Russell DW, Miller AD, Alexander IE (1994) Adeno-associated virus vectors preferentially transduce cells in S phase. Proc Natl Acad Sci USA 91: 8915–8919

Samulski RJ, Berns KI, Tan M, Muzyczka N (1982) Cloning of adeno-associated virus into PBR322: rescue of intact virus from recombinant plasmid in human cells. Proc Natl Acad Sci USA 79: 2077–2081

Samulski RJ, Chang L-S, Shenk T (1989) Helper-free stocks of recombinant adeno-associated viruses: normal integration does not require viral gene expression. J Virol 63: 3822–3828

Samulski RJ, Zhu X, Xiao X, Brook JD, Housman DE, Epstein N, Hunter LA (1991) Targeted integration of adeno-associated virus (AAV) into human chromosome 19. EMBO J 10: 3941–3950

Sullenger BA, Gallardo HF, Ungers GE, Gilboa E (1990) Overexpression of TAR sequences renders cells resistant to human immunodeficiency virus replication. Cell 63: 601–608

Sutherland JH, Lansdorp PM, Kenkelman DH, Eaves AC, Eaves CJ (1990) Functional characterization of individual human hematopoietic stem cells cultured at limiting dilution on supportive marrow stormal layers. Proc Natl Acad Sci USA 87: 3584–3588

Sutherland HJ, Eaves CJ, Lansdorp PM, Thacker JD, Hogge DE (1991) Differential regulation of primitive human hematopoietic cells in long-term cultures maintained on genetically engineered murine stromal cells. Blood 78: 666–672

Walsh CE, Nienhuis AW, Samulski RJ, Brown MG, Miller JL, Young NS, Liu JM (1994) Phenotypic correction of Fanconi anemia in human hematopoietic cells with a recombinant adeno-associated virus vector. J Clin Invest 94: 1440–1448

Wong KK Jr, Rose JA, Chatterjee S (1991) Restriction of HSV-1 production in cell lines transduced with an antisense viral vector targeting the HSV-1 ICP4 gene. Vaccine 91: 183–189

Yu M, Poeschla E, Wong-Staal F (1994) Progress towards gene therapy for HIV infection. Gene Ther 1: 13–26

Zhou SZ, Broxmeyer HE, Cooper S, Harrington M, Srivastava A (1993) Adeno-associated virus 2-mediated gene transfer in murine hematopoietic progenitor cells. Exp Hematol 21: 928–933

Zhou SZ, Cooper S, Kang LY, Ruggieri L, Heimfeld S, Srivastava A, Broxmeyer HE (1994) Adeno-associated virus 2-mediated high efficiency gene transfer into immature and mature subsets of hematopoietic progenitor cells in human umbilical cord blood. J Exp Med 179: 1867–1875

Adeno-associated Virus Vectors for Gene Transfer into Erythroid Cells

D.M. McCarty and R.J. Samulski

Adeno-associated virus (AAV) vectors have the potential to treat a broad range of genetic diseases, among which, those directed toward blood cells currently hold the greatest promise. There are several reasons why the prospect of transducing blood cells merits such extensive interest. First, the target cells are easily accessible. Populations of cells can be removed from the patient, manipulated ex vivo, and subsequently returned to the patient. They can also be enriched for the cell types of interest, cultured, expanded, and transduced with a number of experimental gene therapy vectors. Perhaps more importantly, pluripotent stem cell populations have been identified which can divide and differentiate to repopulate the blood with the different lineages of cells which make up this complex system.

In the following pages we describe studies which demonstrate that recombinant AAV (rAAV) vectors meet at least five specific criteria which are essential to the development of effective gene therapy protocols for diseases such as thalassemia, sickle cell anemia, and Fanconi's anemia. In addition to meeting concerns for general safety and technical feasibility which apply to any viral vector, rAAV has achieved the following objectives: (1) the vector must maintain stable, high level expression of the therapeutic gene; (2) the transduced gene should exhibit the same cell type specificity and regulated expression as the endogenous gene; (3) the vector must efficiently infect and transduce the desired cell population; (4) the transduced gene product must be biochemically

Gene-Therapy Center and Department of Pharmacology, University of North Carolina, Chapel Hill, NC 27599, USA active; and (5) the transduced cell population must exhibit the phenotypic effects of the therapeutic gene product.

1 Regulated Expression in K562 Cells

The first step in the development of a workable gene therapy vector system is to express the therapeutic gene at high levels in response to cell type-specific transcriptional control. Walsh et al. (1992) designed a recombinant virus to test the ability of globin regulatory sequences to operate within the context of AAV vectors in a human erythroleukemia cell line. Previous studies have shown that the K562 cell line responds to hemin induction by maturation into erythroid cells with concomitant increases in levels of γ-globin gene transcription (Rutherford et al. 1979; Papayannopoulou et al. 1988). They have thus provided a model for regulated globin gene expression which has led to the identification and characterization of the *cis* acting regulatory elements of the globin gene clusters. These elements fall into two general groups: the proximal promoter elements associated with each gene of the globin cluster, and the locus control region (LCR) (Tuan et al. 1985; Grosveld et al. 1987). The LCR was originally identified as a series of DNase hypersensitive sites distributed over a 20 kb region located several kb 5′ to the globin gene cluster. An additional hypersensitive site was found downstream from the globin genes. The DNase hypersensitive sites are designated 5′ HS-1 through HS-4 and, 3′ HS-1 for the upstream and downstream sites, respectively. A 225 bp fragment of HS-2, normally situated approximately 10 kb upstream from the β-like globin gene cluster, is sufficient to confer high level, cell type-specific, hemin-inducibility upon the globin promoters (Philipsen et al. 1990).

In order to test the function of the HS-2 element within the context of an integrated rAAV vector, two constructs were prepared. The first contained the human $^{A}\gamma$-globin coding region under the control of its proximal promoter sequences which were, in turn, fused to the HS-2 element of the globin LCR. The second construct was identical except that it lacked the HS-2 element. Additionally, both recombinants carried the gene neo^{R}, conferring G418 resistance, under the separate transcriptional control of the herpesvirus thymidine kinase (TK) promoter. In order to distinguish the rAAV transduced $^{A}\gamma$-globin transcript from the transcripts of the endogenous genes, the recombinant was marked with a 6 bp deletion in the 5′ untranslated region.

Each of the above constructs was used to produce recombinant virus by the standard transfection and complementation procedure (Samulski et al. 1989). K562 cells were then infected with the rAAV and selected for G418 resistance, presumably conferred through integration of the rAAV vector sequences. Pools of resistant colonies were analyzed by Southern blotting to determine the structure and copy number of the proviral genes. The presence of unrearranged proviruses was indicated by vector size restriction fragments in digests using enzymes which cleaved near the ends of the rAAV terminal repeats (TRs). The rAAV containing the HS-2 element consistently integrated as one or two unrearranged copies per cell, as determined by comparison of the hybridization signal to that of the endogenous $^{A}\gamma$-globin genes. In contrast, the rAAV lacking HS-2 gave rise to numerous rearranged copies of the proviral genome. This vector was only 80% of the size of the wild-type AAV genome, suggesting that rAAV was unstable if it was substantially smaller than genome size. This might be due to a preference for packaging genome size DNA.

The transduced K562 cells were used as a model for expression of $^{A}\gamma$-globin in the context of rAAV proviral sequences. Pools of 30 clones were tested for transcription of the transduced $^{A}\gamma$-globin genes with and without hemin induction. The RNase protection analysis employed in this experiment clearly distinguished the transduced from the endogenous transcripts because of the 6 bp deletion in the 5′ untranslated region of the transduced gene. A primer extension assay had also been performed which confirmed that the exogenous gene transcript was initiated at the appropriate site.

In the pools of cells transduced with rAAV containing HS-2, the level of uninduced transcription from the endogenous and the rAAV $^{A}\gamma$-globin genes were similar. Upon hemin induction, comparable increases in transcription were also observed. Thus, the transduced $^{A}\gamma$-globin gene with recombinant promoter and enhancer elements behaved like the endogenous genes. Further, this was the first demonstration of sequences within the context of an integrated rAAV genome responding to cell type-specific signaling pathways.

The K562 cells transduced with the rAAV construct lacking the HS-2 sequence were also tested for transgene expression. In this case, the levels of transcription from the transduced gene were far lower than those of the endogenous genes. Transcripts from the HS-2^{-} proviruses were only detected after hemin induction. This was at least partly due to the fact that a substantial number of the proviral genes had undergone rearrangement. Therefore, a clone was selected which contained an unrearranged copy of the rAAV genome and was tested for expression of the transgene. Again, the level of transcription was far lower than that of the endogenous genes in both induced and uninduced cells. Thus, the HS-2 element appears to have a critical role in directing the expression of the $^{A}\gamma$-globin gene within the context of a rAAV proviral structure.

Individual clones of the cells transduced with the HS-2^{+} vector were also analyzed. In this case, quantitative reverse transcription and amplification by polymerase chain reaction (RT-PCR) were employed to compare transcription levels of the endogenous vs the transduced genes. The transgenes of seven individual clones were consistently transcribed to approximately 40%–50% of the levels of the endogenous genes after correction for copy number. Following hemin induction, the level of transcripts from the transduced genes rose to approximately 85% of the induced levels of the endogenous genes. This demonstrated that the AAV-$^{A}\gamma$-globin constructs were capable of directing the expression of transgenes in amounts comparable to those of the endogenous genes. This was an important observation with respect to the future use of rAAV as a gene therapy vector.

One concern in the use of rAAV or any viral gene therapy vector is the possibility of replication and dissemination of the recombinant virus. In the case of wild-type AAV, rescue from the integrated state occurs almost exclusively in the presence of a suitable helper virus, typically adenovirus. The ability of the transduced cells to generate replicating rAAV in the presence of adenovirus was determined by Southern blot analysis of low molecular weight DNA. The blots were probed with either transgene- or AAV-specific sequences. No signals were generated with either probe, suggesting that the integrated viral vectors were not rescued from the chromosomal DNA by adenovirus. This was as expected because the rAAV vectors did not contain the AAV *rep* gene which is essential for rescue and replication. The absence of AAV protein coding sequences in the transduced cells demonstrated that the vector preparation was not contaminated with wild-type virus. The potential for regeneration of wild-type AAV in a vector preparation was a concern because of the possibility of recombination between the viral vector and AAV genes contained in the plasmid used to complement for replication and encapsidation. This and other rAAV vectors are, however, designed such that there is no sequence overlap between the vector and the complementing plasmid, thus minimizing the risk of recombination events (SAMULSKI et al. 1989).

2 Functional Regulatory Sequences Within the Adeno-associated Virus Transduced $^{A}\gamma$-Globin

The ability of an AAV-based vector to stably transduce a population of cells and to respond to cell type-specific stimuli makes it a powerful tool for the study of the behaviour of transgenes and associated regulatory elements. MILLER et al. (1993) used a mutational analysis to investigate the roles of the *cis* active sequences within the HS-2 element in the context of rAAV. In addition to the vectors discussed above, which contained the $^{A}\gamma$-globin gene and either a wild-type HS-2 or no LCR element, two new vectors with mutations in HS-2 were created. These contained specific changes in *cis* active elements which rendered them inactive for binding to their cognate transcription factors. Multiple transcriptional activator binding sites are located within the 300–400 bp HS-2 region. The erythroid cell-specific transcription factor NF-E2 (ANDREWS et al. 1993) binds within an enhancer element of HS-2 containing tandem AP-1 sites (NEY et al.1990a,b). The integrity of these sites is essential for hemin inducibility of globin gene expression (SORRENTINO et al. 1990). A second erythroid-specific transcription factor, GATA-1 (PEVNY et al. 1991), also called NFE-1 (YAMAMOTO et al. 1990), binds to a

5′-GATA-3′ consensus motif within HS-2 and is essential for maturation of erythroid cells in transgenic mice. The NF-E2 site was inactivated using two single-base substitutions, the GATA-1 site, by three base changes in each of the symmetric half-sites. These mutants were then tested for their ability to express high levels of $^{A}\gamma$-globin RNA in response to hemin induction.

The erythroid cell line K562 was infected with each of these viral vectors. In order to compare the ability of each enhancer variant to express the $^{A}\gamma$-globin gene upon induction, it was essential that each of the rAAV transgenes was in a similar proviral structure. This being the case, differences in expression could be attributed to the effects of the specific *cis* active sequences associated with the transgene or to position effects related to the site of chromosomal integration. Position effects were evaluated by comparison of clonal cells containing the same rAAV construct but resulting from a separate transduction event in a presumably different location within the genome. Ten G418 resistant colonies were isolated from each of the viral vector infections. Southern blots of genomic DNA from individual clones were analyzed to determine the copy number, structure, and integration status of the recombinant proviruses.

Three of the vector constructs, containing the wild-type HS-2 region, the mutant NFE-2 site, or the mutant GATA-1 site, yielded similar patterns of integration. Half of the 30 clones analyzed (10 from each vector) contained rAAV which integrated as a single unrearranged copy. In contrast, the construct lacking the HS-2 region yielded a higher number (60%) of rearranged integrated vectors as had been seen previously. None of the recombinant virus sequences in these cells appeared to be episomal because no free ends were detected in these restriction digests.

The transduced cell lines were subsequently analyzed for expression from the exogenous gene at the level of RNA transcription. Several clones were chosen, each containing a single, unrearranged proviral copy of the transgene. Messenger RNA from either hemin-induced or uninduced cells was subjected to RT-PCR. The level of expression from the transgenes with their various *cis* elements was determined relative to the endogenous γ-globin genes present in the K562 cell line. Cell lines containing each of the four promoter constructs were compared with and without hemin induction.

In the absence of hemin induction, the proviral $^{A}\gamma$-globin gene under the control of the NF-E2 binding site mutant was expressed at a comparable level to that of the transgene with the wild-type HS-2. Both were higher than that of the construct lacking the HS-2 region. However, in spite of its ability to express $^{A}\gamma$-globin at wild-type basal levels, the NF-E2 mutant could not be induced to higher levels of expression upon hemin induction.

In contrast, the clones containing proviral $^{A}\gamma$-globin under the control of the GATA-1 mutant were induced to levels comparable to those containing the wild-type HS-2 region. Interestingly, Miller et al. noted a higher degree of variability in the level of induced and uninduced expression from the GATA-1 mutants. This suggested that, although GATA-1 was not essential for either basal or induced expression, the ability to bind this factor mitigated the effects of the site of integration, which is presumably the only difference between these cells clones. All of the clones had been selected for expression of the TK-*neo* gene which was also present on the recombinant viral genome. It is thus implicit that each of these clones had the provirus integrated into an actively transcribed region. The GATA-1 dependence for consistent high level expression suggests that the chromosomal context is not an all or none mediator of transcriptional potential.

Taken together, MILLER et al. have succeeded in demonstrating two important points about AAV-derived gene therapy vectors. First, the exogenous $^{A}\gamma$-globin gene under the control of its core promoter and a small segment of the globin LCR was able to express significant levels of globin message and was inducible with hemin. Thus, within the context of the AAV provirus, this gene behaved indistinguishably from the endogenous gene in these cells. This represents an important foundation for the establishment of the utility of rAAV vectors in achieving normal control of expression from a transduced gene. Secondly, as long as the proviral construct contained the appropriate sequences (GATA-1), the expression of the transgene was independent of chromosomal position. This is an essential feature for these vectors in light of the recent observation that AAV integration may be random in the absence of the viral *rep* gene product (WALSH et al. 1992; RUSSELL et al. 1994).

The success of this series of experiments suggests a format for the construction of a rAAV vector for the treatment of hemoglobinopathies. It is clear that elements of the globin LCR will be essential to high level, regulated expression of transduced globin genes. In a therapeutic vector, it will not necessarily be desirable to maintain the *neo*^{R} marker within the vector genome. Based upon this study, however, additional sequences will be required simply to achieve a genome of approximately wild-type size in order to ensure the stability of the vector. These could be comprised of additional sequences of the LCR which may confer yet higher levels of transduced globin gene expression. While the HS-2 element is sufficient to confer transcription levels comparable to those of the endogenous genes in K562 cells, this might not be the case in other erythroid cell lines or in primary hematopoietic cells. Footprinting studies in vivo have revealed that, while the HS-2 transcription factor binding sites were occupied in K562 cells, the cognate sites of the HS-3 and the $^{A}\gamma 3'$ enhancer were not bound (IKUTA and KAN 1991; STRAUSS et al. 1992). This was the case whether or not the cells were induced with hemin. It is therefore possible that the behaviour of the rAAV vector construct in K562 cells does not reflect its ability to direct endogenous levels of γ-globin transcription in other erythroid cell types where all of the LCR elements may be engaged in the activation of globin gene expression.

In addition to the direct applicability of these studies to the design of gene therapy protocols for hemoglobinopathies, this work also illustrates the power of the rAAV transduction system as a tool for genetic research. The authors have described systems wherein the *cis* active elements associated with the rAAV transgene can be manipulated to determine the optimal combinations required for regulated gene expression.

Studies of this type can be used to investigate the role of any potential *cis* acting sequences which can be accommodated within the vector containing a linked reporter gene. In cell types in which AAV infection is relatively efficient, the transgene expression can be characterized without prior selection or establishment of cloned cell lines. The application of rAAV vectors for research into the behaviour of genetic components within the context of the cellular chromosome will allow rapid advances in the understanding of eukaryotic gene expression.

3 Recombinant Adeno-associated Virus Transduction of Hematopoietic Progenitor Cells

One of the most important features of any proposed gene therapy vector is its ability to target the population of cells which either manifest the genetic defect or give rise to that population. The ability of a vector to transduce the hematopoietic stem cell population will therefore determine its utility for the treatment of diseases arising from defects in blood cell populations. GOODMAN et al. (1994) have successfully demonstrated that rAAV vectors can infect primary hematopoietic progenitor cells and maintain stable expression of a transduced gene. In this case a rAAV vector expressing the *E. coli* β-galactosidase gene (AAV-β-gal) was used to investigate the infectivity of these vectors in stem cells in vitro.

The AAV-β-gal vector was initially titered and characterized in Detroit 6 cells in the absence of a selectable marker. The duration of β-gal expression in these cells was evaluated by infecting the cells and passing them twice without selection. Interestingly, the number of β-gal positive cells declined between the first and seventh day postinfection. The remaining positive cells were grouped in clusters, suggesting that they were derived from a single cell carrying an integrated rAAV genome. The reason for the initial decline in cells expressing β-gal may have been the loss of episomal rAAV genomes in successive cell divisions. This suggested that the rate of rAAV infection exceeded the rate of integration in these cells. Studies by other investigators have led to similar conclusions and this may be a general trait of rAAV vectors (FLOTTE et al. 1994).

The ability of the rAAV vector to infect and be expressed in the cells of the hematopoietic lineage was evaluated on several different cell types. First, the AAV-β-gal vector was used to transduce K562 cells. As discussed above, the rAAV vector expressing a human $^{A}\gamma$-globin gene and the selectable *neo*^{R} gene previously had been used to infect this cell type. In this case, however, the efficiency of infection could be measured directly, by X-gal staining, and no selection pressure was applied. GOODMAN et al. found that the rate of infection in the K562 cells was only 2%–3% of that observed in Detroit 6 cells. The relative efficiency of subsequent integration in K562 cells was not evaluated.

Goodman et al. next determined the utility of this vector in transducing primary hematopoietic cells. Normal human bone marrow cells were obtained and low density mononuclear cells were isolated through buoyant density centrifugation. This population was subsequently enriched for cells expressing the CD34 antigen, a marker for pluripotent stem cells, by immunoselection. The $CD34^+$ cells were infected with rAAV vectors by exposure to the virus at a multiplicity of infection (MOI) between 1 and 10 over a period of 3 days. Approximately 60%–70% of the $CD34^+$ enriched cells stained positive for β-gal. This demonstrated the ability of rAAV vectors to efficiently infect primary hematopoietic cells and express an exogenous gene.

Subsequent experiments were designed to demonstrate that the transduced cells were indeed hematopoietic progenitors and the rAAV genome was stable. This was evaluated by infecting $CD34^+$ cells with AAV-β-gal and allowing the cells to form colonies under methylcellulose. The transduced state of the resulting colonies was determined using a PCR-based assay rather than histochemical staining. The evaluation of the extent of β-gal expression in the intact colonies had proven difficult due to high background staining. A cell line previously determined to carry a single integrated copy of the AAV-β-gal vector was used to calibrate the system such that it was sensitive to the copy number of the vector over a broad range of genomic DNA concentrations. The copy number of the vector relative to the cellular genome was determined using primers specific for the cellular β-actin gene to produce a standard for comparison.

The infections of the $CD34^+$ enriched cells were performed at approximately MOI 1, as determined on Detroit-6 cells. These were subsequently plated under methylcellulose without selection. Individual colonies were picked between 12 and 14 days postinfection. Approximately 60% of the colonies (7 of 12) yielded PCR signals indicating the presence of the *lacZ* sequence. The signals were variable but comparable in intensity to that of the β-actin control. This suggested that the vector DNA had integrated at a copy number in the range of one to two per cell.

Similar experiments were performed with $CD34^+$ progenitor cells from rhesus monkeys. A transduction frequency of 66% (10 of 15) was obtained in the monkey hematopoietic cells. Half of the positive colonies yielded *LacZ*-specific PCR signals which were approximately 50% of the intensity of the γ-globin standard while the other half yielded signals of similar intensity to the standard. As in the human $CD34^+$ cells, this suggested that the vector copy numbers were approximately one and two per cell. Taken together, the integration patterns in primary hematopoietic progenitor cells were consistent with those previously observed in the γ-globin transduced K562 cells.

Wild-type AAV exhibits a preference for integration into a small region of human chromosome 19. The ability of AAV, or recombinants containing the *rep* gene, to target this site in primary cells, as opposed to established cell lines, had not previously been tested. Goodman et al. used an adaptation of the PCR-based assay described above to determine whether the wild-type AAV virus would preferentially integrate into the chromosome 19 site. In this case, nested sets of primers specific for the vector sequence and for the chromosome 19 integration site were used to amplify potential junctions between the proviral and cellular DNA.

Wild-type AAV was used to infect human primary hematopoietic cells at a high multiplicity. Individual colonies derived from these infected cells were first tested for the presence of any AAV-specific sequences using PCR primers internal to the AAV capsid gene. Approximately 70%–89% of the colonies were positive for AAV sequences, although they might have been amplified from episomal as well as integrated copies of the AAV DNA. Pools of infected cells, derived from 40-80 individual colonies, were then tested for the presence of junction sequences which would indicate site-specific integration. Using the AAV and chromosome 19 primers, each pool yielded PCR signals consistent with the presence of AAV-chromosome 19 junctions. The amplified DNA from one of the pools was cloned into a bacterial plasmid for analysis of the junction sequences. This revealed an AAV-chromosome 19 junction which was similar in structure to previously characterized junctions sequences (Kotin et al. 1992). The breakpoint within the chromosome 19 preintegration site was approximately 335 bp from a sequence known to be capable of interacting with AAV Rep protein in vitro (Weitzman et al. 1994). The breakpoint in the AAV sequence occurred within the TR, as expected. Typical of these junction compositions, the TR sequence was incomplete and had suffered truncation within one of the small internal palindromes which form the T-shaped structure of the folded TR. This is consistent with previous analysis of AAV provirus structures and, indeed, no intact TR sequence has yet been found as part of a chromosomal junction. Presumably, AAV relies on intact copies of the TR within a tandem array of provirus for rescue and replication of the viral genome. Cell lines containing multiple copies of the AAV genome integrated as a tandem array have generally been observed to be the most efficient at subsequent rescue from the latent state (Laughlin et al. 1986; McLaughlin et al. 1988).

The integration sites in a number of cloned colonies were also analyzed using the nested primer set specific for AAV-chromosome 19 junctions. Four of 27 individual colonies yielded PCR signals consistent with specific junctions. Three of these were subsequently cloned and sequenced and each was found to contain a novel junction. Thus, the wild-type AAV is capable of targeted integration in hematopoietic progenitor cells with a minimum frequency of 14%. Because the first AAV-specific primer of the nested set annealed to a sequence unique to the right hand end of the viral genome, this assay would not have detected provirus structures in the opposite orientation with respect to the chromosome 19 sequence. It is therefore possible that the real frequency of targeted integration is at least twice this minimum.

4 Biological Activity of the Transduced γ-Globin Gene Product

In a study aimed at a potential gene therapy protocol for the treatment of sickle cell anemia, Miller et al. (1994) have used the transduction of hematopoietic progenitor cells with rAAV-γ-globin to address the efficacy of rAAV vectors. The authors utilized target cells from a sickle cell anemia patient such that the fate of the transduced gene product could be followed in terms of its association with components of biologically active complexes. The rAAV contained, in addition to a marked γ-globin gene with a 6 bp deletion in the 5′ untranslated region, sequences of the HS-4, HS-3, HS-2 regions of the globin LCR. There was no selectable marker included in this vector. The rAAV was grown in the conventional manner and the particle number was estimated using quantitative PCR to measure the amount of DNase resistant vector-specific DNA. This represented the number of encapsidated rAAV genomes which, in turn, could be related to the number of infectious units of viral vector.

Human peripheral blood mononuclear cells from a sickle cell anemia patient were selected for expression of the CD34 antigen and infected with the rAAV at a ratio of 500–1000 particles per burst forming units-erythroid (BFU-E). This corresponded to roughly five to ten infectious units per cell. After exposure to the rAAV-γ-globin vector, the $CD34^+$ cells were plated in methylcellulose containing growth factors and incubated for 13–19 days.

The high particle to cell ratio used to infect the $CD34^+$ cells led to substantial background levels when the quantitative PCR assay was employed to determine the infection efficiency. Therefore, the rate of transduction was estimated using RT-PCR to measure the expression of the rAAV-γ-globin gene. One of the primers spanned the 6 bp deletion in the leader sequence of the transduced γ-globin gene and did not amplify transcripts from the endogenous gene. The signals that were generated from amplification of the vector DNA were distinguishable by size from the signals derived from its transcripts because the primer pair spanned an intron sequence.

Nucleic acid from individual colonies was analyzed for expression of the transduced gene. Some 20%–30% of the erythroid colonies yielded positive RT-PCR signals. Considering the ratio of approximately 5–10 infectious units per cell, this suggested that the efficiency of AAV infection in this cell type was relatively low. However, the infectious unit titer of this vector preparation could have been lower than was estimated based on the total particle number. Nonerythroid colonies did not yield RT-PCR signals with these primers, suggesting that the *cis* acting elements of the γ-globin promoter and the globin LCR were sufficient to confer cell type-specific expression.

The RT-PCR assay was also used to determine the level of expression of the transduced gene compared to the endogenous γ-globin. The endogenous transcripts from the rAAV positive colonies were amplified using a primer which included sequences that had been deleted from the rAAV-γ-globin gene. These comparisons suggested that the γ-globin of the rAAV vector gave rise to transcript levels ranging from 4% to 71% of the endogenous gene. Neither the copy number nor the integration status of the rAAV vectors was investigated in cells from these colonies. Therefore, differences in the copy number of integrated proviral genomes or, alternatively, expression from episomal rAAV might have accounted for some of this variability.

The most interesting aspect of this study was the subsequent analysis of the globin gene products in transduced and nontransduced cell pools derived from sickle cell. The hemoglobin compositions of erythroblasts from these pools were chromatographed using an HPLC column. Hemoglobin complexes from the mock infected cells were composed of 26% fetal globin and 70% sickle cell globin. In a pool of cells derived from BFU-E which had been exposed to rAAV-γ-globin, fetal hemoglobin constituted 40% of the total. These values probably reflect an even greater increase in the fetal hemoglobin content of the transduced cells because they comprised only approximately 50% of the erythroblast pools.

This observation has promising implications for the future of gene therapy for sickle cell anemia. The presence of elevated levels of fetal hemoglobin within red blood cell leads to the formation of functional complexes composed of α-globin and γ-globin preferentially over α-globin and sickle cell globin. This reduces the tendency of the sickle globins to polymerize and thus ameliorates the disease phenotype.

5 Phenotypic Correction of a Genetic Disease In Vitro Using a Recombinant Adeno-associated Virus Transduced Gene

The studies presented thus far have demonstrated that rAAV vectors can transduce hematopoietic cell populations and express a gene product with biological activity. The remaining problem of demonstrating a phenotypic effect from the gene therapy vector has been addressed using a different disease model, Fanconi's anemia (FA) (Liu et al. 1994; Walsh et al. 1994). The feature of this disease which makes it a particularly useful model is that the genetic defect results in aberrations which can be observed at the level of a single cell. The disease is an autosomal recessive disorder which appears to stem from deficiencies in DNA repair. There are three known FA complementation groups (A, B, and C), all of which give rise to similar clinical syndromes. Manifestations of the disease include physical anomalies such as shortened digits, pancytopenia, and a susceptibility to leukemia and solid tumors (Butturini et al. 1994). The latter characteristic is probably the result of gross chromosomal rearrangements which are a hallmark of the disease at the cellular level. The chromosomal aberrations are far more pronounced when cells from a FA patient are cultivated in the presence of mitomycin C, a DNA cross-linking reagent. Fanconi's anemia cells are more than 100-fold more sensitive than normal cells to mitomycin C.

One of the three FA complementation group genes, group C (FACC), has been cloned and mapped to a specific location on chromosome 9q of the human genome (GIBSON et al. 1994). This gene encoded a 63 kDa protein and can correct the phenotypic defect when transfected into FA cells in culture. Phenotypic correction is easily assessed by cell survival and by the absence of chromosomal aberrations after treatment with mitomycin C.

The primary treatments for this disorder have been aimed at the manifestations of the disease related to blood cell disorders such as anemia and leukemia. It will therefore also serve as a useful model for gene therapy targeted toward hematopoietic cells. A rAAV vector was created with the FACC cDNA expressed from the Rous sarcoma virus (RSV) long terminal repeat (LTR) promoter. The FACC coding sequence was small enough that the *neo*^{R} gene with a separate TK promoter could be accommodated within the rAAV as a selectable marker. This allowed the recombinant virus to be titered by transduction to G418 resistance in Detroit 6 cells.

The AAV-FA vectors were initially tested for phenotypic correction on two Epstein-Barr virus-transformed lymphoblastic cell lines derived from FA patients. The cells were infected with the rAAV at a multiplicity of approximately 0.3 per cell. Infected cells were then selected using the *neo*^{R} marker. The structure of the rAAV-FA vector within the G418 resistant pools was characterized by Southern blot analysis. Genomic DNA was digested with *SnaB*I which cuts within each TR of AAV. Hybridization of the blots with a probe specific for the *neo*^{R} sequences revealed a single fragment of the predicted length for intact AAV vector, indicating that the *neo*^{R} and FACC were being expressed from intact, unrearranged copies of the vector genome.

An RT-PCR assay was used to compare the level of expression from the FACC transgene relative to that of the endogenous gene. The transduced and the endogenous FACC genes each had unique 3′ untranslated regions which were used to amplify specific products from the respective transcripts. Total cellular RNAs from both the AAV-FACC transduced and parental (BD0215) lymphoblastic cell lines were reverse transcribed and amplified with each primer pair. Amplification products of the predicted sizes for both of the genes were generated from the transduced cell lines whereas only the endogenous gene transcript was observed in the parental FA cells.

The expression of the FACC transgene was further characterized by Immunoprecipitation of the exogenous gene product. An anti-FACC antiserum, which was specific for the COOH-terminal one third protein, was used in immunoprecipitation reactions with [^{35}S] methionine-labeled cell lysates. The mutation in BD0215 was a nonsense codon creating a truncated gene product which was not recognized by this antiserum. As expected, no specific proteins were precipitated from the parental BD0215 lymphoblasts. In contrast, the rAAV-FACC transduced cells produced a protein which comigrated with the wild-type gene product from normal lymphoblasts. The transduced and FACC wild-type cells produced similar amounts of protein, suggesting that the level of transcription from the RSV promoter would be sufficient for a therapeutic effect.

The phenotypes of the transduced FA cells were subsequently characterized with respect to several parameters reflecting the DNA repair defect. Pools of transduced cells were exposed to concentrations of mitomycin C, ranging from 1 to 1000 n*M*, for a period of 5 days. Transduction of two different FA lymphoblastic cell lines (BD0215 and HSC536) resulted in approximately 100-fold increases in resistance to the DNA cross-linking reagent. This represented an increase in EC_{50} to a level comparable to that of normal lymphoblastic cells.

The mitomycin C sensitivity FA cells can also be observed at the cytogenic level. Exposure to the drug induces high levels of chromosomal breakage and radial formation, presumably due to the inability to repair cross-linked DNA. The extent of chromosomal aberration is easily observed in cytogenetic assays. The rAAV-FACC transduced cells suffered approximately ten fold lower levels of chromosomal aberrations than the parental cells exposed to the same concentration of mitomycin C. This assay was performed on cells which had undergone greater than 50 serial passages in the absence of *neo*^{R} selection pressure. It was therefore likely that each cell had stably integrated the rAAV vector.

A third assay which distinguished FA from normal cells relied on flow cytometric analysis to measure the distribution of cells within the cell cycle. FA cells have been observed to accumulate in the G2 phase and this tendency is more pronounced in the presence of mitomycin C. The parental FA lymphoblastic cell lines used in this study displayed this pattern. In contrast, the pool of rAAV-FACC-transduced FA lymphocytes displayed flow histograms which were characteristic of normal cells. Thus, the transduced cell population is able to proceed through the cell cycle without the abnormally long pause in the G2 phase.

The normal cycling of the transduced lymphoblast population has important implications for the efficacy of gene therapy treatments for FA. Because the transduced cells cycle faster, and therefore grow faster, than the nontransduced population, a relatively small fraction of phenotypically corrected stem cells could be expected to make a significant contribution to the population of circulating blood cells invivo. A treatment regimen for FA might therefore require only that an aliquot of hematopoietic stem cells be transduced ex vivo and reintroduced into the patient. It would not be necessary to ablate the patient's untreated bone marrow cells prior to reintroduction because the corrected cells would maintain a growth advantage. These experiments also demonstrate the lack of toxicity from the expression of the transduced FACC gene in an unregulated manner from a viral promoter, suggesting that it would be a suitable construct for gene therapy.

The ability of the rAAV-FA vector to infect, transduce, and phenotypically correct hematopoietic stem cells was assessed using CD34 selected cells. A population composed of 92% $CD34^+$ was prepared from the peripheral blood mononuclear cells of an FA patient. The cells were exposed to the AAV-FA vector at a MOI of 0.1 per cell for a period of 3 days. Subsequently, infected and mock-infected CD34$^+$ cells were plated in methylcellulose and the colonies were counted after 15 days in culture. The rAAV-infected cells gave rise to four fold more colonies than the mock-infected cells. This probably reflects the growth advantage of the FACC corrected phenotype. Infected and mock-infected cells were also plated in methylcellulose containing 1 n*M* mitomycin C. Under these conditions, the transduced cells gave rise to eight- to ten-fold more colonies than the mock-infected. Thus, cells which were capable of reconstituting bone marrow and renewing blood cell populations could be transduced to a wild-type phenotype using the AAV-FACC vector. Further, even in the absence of selection pressure from mitomycin C, the transduced cells displayed a substantial growth advantage over the nontransduced cells.

When colonies from the rAAV-FACC infected cells were subsequently picked and tested by RT-PCR for expression of the exogenous gene, only 60% yielded positive signals. This was in spite of the fact that treatment with rAAV-FACC resulted in an eight to ten fold increase in the number of resistant colonies. This observation raises the question of whether the vector was stably integrated in all of the progenitor cells. It was possible that FACC was expressed from episomal copies of the vector in the early stages of colony formation, thus providing sufficient quantities of the gene product to overcome the inhibition of cell cycling. After successive cell divisions, however, the episomes may have been lost and the FACC gene products diluted out.

The issue of the rate of integration versus the rate of infection and transduction will be an important focus of research in rAAV vectors. Already, recent studies have demonstrated that a variety of stress inducing treatments such as heat shock and radiation as well as specific adenovirus gene products will dramatically increase the rate of transduction with rAAV in cultured cells (Alexander et al. 1994; Russel et al. 1995; Ferrari and Samulski, unpublished). It is likely that the common mechanism in these disparate treatments is the promotion of synthesis of the complementary strand of the single-stranded virion DNA. Unlike retrovirus vectors, which cannot express the transduced gene product without first having integrated into the chromosome, rAAV vectors appear to be capable of transient expression prior to integration. This is perhaps not surprising considering the biology of the virus. Wild-type AAV is not known to have any nonstructural proteins associated with the virion (although this point is difficult to formally exclude). If indeed the *rep* gene is involved at some step in the integration process, it must be expressed from the episomal genome prior to integration.

The integration of rAAV vectors should perhaps be considered an aberration of the normal process of AAV integration because the *rep* gene is not present. We have already noted two differences in the integration phenotype of *rep*$^-$ AAV. First, they have a higher rate of transduction then the *rep*$^+$ vectors. Second, they do not appear to exhibit the preference for integration into the characteristic site on chromosome 19. Nonetheless, there have not been any differences found between the structures of the junction sequences of *rep*$^+$ and *rep*$^-$ provirus. It is therefore likely that the integration is mediated by a cell-specific mechanism in both cases and that a *rep* gene product is capable of interacting with and modulating the process. A clearer understanding of the AAV integration mechanism will undoubtedly provide a great deal of insight into the cell growth conditions which will engender efficient and permanent transduction rather than transient expression from an episome.

6 Conclusion

The studies to date represent a systematic approach to assessing AAV vectors for their potential to transduce a therapeutic gene (globin, FACC) into the appropriate target tissue (erythroid, bone marrow stem cells). While the supporting data suggest that AAV will carry gene cassettes such as the globin or FACC gene, regulated by tissue-specific or viral-specific promoters, and generate the appropriate therapeutic message, these studies will be incomplete until addressed in vivo. Because the long term success of high level expression of transduced genes from rAAV vectors will ultimately have to be evaluated using in vivo models, vector titers also become an issue of importance. Currently, only pilot scale studies involving single animals have been attempted. The major reason for this shortcoming is the difficulty in generating sufficient amounts of the vectors for administration into animals. While studies in tissue culture cell lines have clearly addressed the fate of the transduced gene, this question remains uncertain in the bone marrow stem cell target. It therefor appears that, coupled with in vivo studies, the ability to test for vector integration will help assess the final utility of AAV vectors for gene therapy of hemoglobinopathies. In continuing studies, in which small areas of tissue can be targeted for transduction with rAAV vectors, such as brain (Kaplitt et al. 1994) and muscle (Xiao and Samulski, unpublished), the duration of transgene expression and the absence of pathology at the injection sites have been encouraging. Overall, the nonpathogenic AAV vectors may fulfill a critical niche in gene transfer for the correction of genetic disorders.

References

Alexander IE, Russell DW, Miller AD (1994) DNA-damaging agents greatly increase the transduction of nondividing cells by adeno-associated virus vectors. J Virol 68(12): 8282–8287

Andrews NC, Erdjument-Bromage H, Davidson MB, Tempst P, Orkin SH (1993) Erythroid transcription factor NF-E2 is a haematopoietic-specific basic-leucine zipper protein. Nature 362(6422): 722–728

Butturini A, Gale RP, Verlander PC, Adler-Brecher B, Gillio AP, Auerbach AD (1994) Hematologic abnormalities in Fanconi anemia: an International Fanconi Anemia Registry study. Blood 84(5): 1650–1655

Flotte TR, Afione SA, Zeitlin PL (1994) Adeno-associated virus vector gene expression occurs in nondividing cells in the absence of vector DNA integration. Am J Respir Cell Mol Biol 11(5): 517–521

Gibson RA, Ford D, Jansen S, Savoia A, Havenga C, Milner RD, de-Ravel TJ, Cohn RJ, Ball SE, Roberts I et al (1994) Genetic mapping of the FACC gene and linkage analysis in Fanconi anaemia families. J Med Genet 31(11): 868–871
Goodman S, Xiao X, Donahue RE, Moulton A, Miller J, Walsh C, Young NS, Samulski RJ, Nienhuis AW (1994) Recombinant adeno-associated virus mediated gene transfer into hematopoietic progenitor cells. Blood 84(5): 1492–1500
Grosveld F, van-Assendelft GB, Greaves DR, Kollias G (1987) Position independent, high-level expression of the human beta-globin gene in transgenic mice. Cell 51(6): 975–985
Ikuta T, Kan YW (1991) In vivo protein-DNA interactions at the beta-globin gene locus. Proc Natl Acad Sci USA 88(22): 10188–10192
Kaplitt MG, Leone P, Samulski RJ, Xiao X, Pfa DW, O'Malley KL, During MJ (1994) Long-term gene expression and phenotypic correction using adeno-associated virus vectors in the mammalian brain. Nature Genet 8: 148–154
Kotin RM, Linden RM, Berns KL (1992) Characterization of a preferred site on human chromosome 19q for integration of adeno-associated virus DNA by non-homologous recombination. EMBO J 11(13): 5071–5078
Laughlin CA, Cardellichio CB, Coon HC (1986) Latent infection of KB cells with adeno-associated virus type 2. J Virol 60: 515–524
Liu JM, Buchwald M, Walsh CE, Young NS (1994) Fanconi anemia and novel strategies for therapy. Blood 84(12): 3995–4007
McLaughlin SK, Collis P, Hermonat PL, Muzyczka N (1988) Adeno-associated virus general transduction vectors: analysis of proviral structures. J Virol 62(6): 1963–1973
Miller JL, Walsh CE, Ney PA, Samulski RJ, Nienhuis AW (1993) Single-copy transduction and expression of human gamma-globin in K562 erythroleukemia cells using recombinant adeno-associated virus vectors: the eect of mutations in NF-E2 and GATA-1 binding motifs within the hypersensitivity site 2 enhancer. Blood 82(6): 1900–1906
Miller JL, Donahue RE, Sellers SE, Samulski RJ, Young NS, Nienhuis AW (1994) Recombinant adeno-associated virus (rAAV)-mediated expression of a human gamma-globin gene in human progenitor-derived erythroid cells. Proc Natl Acad Sci USA 91(21): 10183–10187
Ney PA, Sorrentino BP, Lowrey CH, Nienhuis AW (1990a) Inducibility of the HS II enhancer depends on binding of an erythroid specific nuclear protein. Nucleic Acids Res 18(20): 6011–6017
Ney PA, Sorrentino BP, McDonagh KT, Nienhuis AW (1990b) Tandem AP-1-binding sites within the human beta-globin dominant control region function as an inducible enhancer in erythroid cells. Genes Dev 4(6): 993–1006
Papayannopoulou T, Nakamoto B, Kurachi S, Tweeddale M, Messner H (1988) Surface antigenic profile and globin phenotype of two new human erythroleukemia lines: characterization and interpretations. Blood 72(3): 1029–1038
Pevny L, Simon MC, Robertson E, Klein WH, Tsai SF, Dagati V, Orkin SH, Costantini F (1991) Erythroid dierentiation in chimaeric mice blocked by a targeted mutation in the gene for transcription factor GATA-1. Nature 349(6306): 257–260
Philipsen S, Talbot D, Fraser P, Grosveld F (1990) The beta-globin dominant control region: hypersensitive site 2. EMBO J 9(7): 2159–2167
Russell DW, Miller AD, Alexander IE (1994) Adeno-associated virus vectors preferentially transduce cells in S phase. Proc Natl Acad Sci USA 91(19): 8915–8919
Russell DW, Alexander IA, Miller AD (1995) DNA synthesis and topoisomerase inhibitors increase transduction by adeno-associated virus vectors. Proc Natl Acad Sci USA 92: 5719–5723
Rutherford TR, Clegg JB, Weatherall DJ (1979) K562 Human leukaemic cells synthesize embryonic haemoglobin in response to haemin. Nature 280: 164–165
Samulski RJ, Chang LS, Shenk T (1989) Helper-free stocks of recombinant adeno-associated viruses: normal integration does not require viral gene expression. J Virol 63: 3822–3828
Sorrentino B, Ney P, Bodine D, Nienhius AW (1990) A 46 base pair enhancer sequence within the locus activating region is required for induced expression of the gamma-globin gene during erythroid dierentiation. Nucleic Acids Res 18(9): 2721–2731
Strauss EC, Andrews NC, Higgs DR, Orkin SH (1992) In vivo footprinting of the human alpha-globin locus upstream regulatory element by guanine and adenine ligation-mediated polymerase chain reaction. Mol Cell Biol 12(5): 2135–2142
Tuan D, Solomon W, Li Q, London IM (1985) The "beta-like-globin" gene domain in human erythroid cells. Proc Natl Acad Sci USA 82(19): 6384–6388

Walsh CE, Liu JM, Xiao X, Young NS, Nienhuis AW, Samulski RJ (1992) Regulated high level expression of a human gamma-globin gene introduced into erythroid cells by an adeno-associated virus vector. Proc Natl Acad Sci USA 89(15): 7257–7261

Walsh CE, Nienhuis AW, Samulski RJ, Brown MG, Miller JL, Young NS, Liu JM (1994) Phenotypic correction of Fanconi anemia in human hematopoietic cells with a recombinant adeno-associated virus vector (see comments). J Clin Invest 94(4): 1440–1448

Weitzman MD, Kyostio SR, Kotin RM, Owens RA (1994) Adeno-associated virus (AAV) Rep proteins mediate complex formation between AAV DNA and its integration site in human DNA. Proc Natl Acad Sci USA 91(13): 5808–5812

Yamamoto M, Ko LJ, Leonard MW, Beug H, Orkin SH, Engel JD (1990) Activity and tissue-specific expression of the transcription factor NF-E1 multigene family. Genes Dev 4(10): 1650–1662

Adeno-associated Virus 2-Mediated Transduction and Erythroid Lineage-Specific Expression in Human Hematopoietic Progenitor Cells

A. Srivastava[1], X.-S. Wang[2], S. Ponnazhagan[2], S.Z. Zhou[2], and M.C. Yoder[3]

1 Introduction

Parvoviruses are among the smallest of the DNA-containing viruses that infect a wide variety of vertebrates (Siegl et al. 1985). Two parvoviruses of human origin, the nonpathogenic adeno-associated virus 2 (AAV) and the parvovirus B19, a common human pathogen, have been studied extensively (Berns and Bohenzky

[1]Division of Hematology and Oncology, Departments of Medicine, and Microbiology and Immunology, and the Walther Oncology Center, Indiana University School of Medicine, Indianapolis, IN 46202–5120, USA
[2]Division of Hematology and Oncology, Department of Medicine and the Walther Oncology Center, Indiana University School of Medicine, Indianapolis, IN 46202–5120, USA
[3]Departments of Pediatrics, and Biochemistry and Molecular Biology and the Herman B Wells Center, Indiana University School of Medicine, Indianapolis, IN 46202-5120, USA

1987; Brown et al. 1994). AAV requires coinfection with a helper virus, such as adenovirus or herpesvirus, for its optimal replication (Berns 1990), but in the absence of a helper virus, the AAV genome establishes a latent infection in a site-specific manner (Kotin and Berns 1989; Kotin et al. 1990, 1991, 1992; Samulski et al. 1991). B19, by contrast, is an autonomously replicating virus with a remarkable tropism for human erythroid progenitor cells (Ozawa et al. 1986, 1987; Yaegashi et al. 1989; Srivastava and Lu 1988; Takahashi et al. 1990). We have described the construction of a recombinant AAV-B19 hybrid genome, in which we combined the remarkable features of these two parvoviruses, and speculated that such a hybrid vector may prove useful for high efficiency transduction of primary human hematopoietic progenitor cells (Srivastava et al. 1989). Indeed, it has become increasingly clear that the AAV-based vector system may prove to be a useful alternative to the more commonly used retroviral and adenoviral vectors for its potential use in human gene therapy (Muzyczka 1992; Carter 1993; Srivastava 1994). Despite these advances, a number of fundamental questions related to AAV remain unanswered. For example, the molecular details of viral assembly and the mechanism of viral entry into the host cell have not been rigorously analyzed. Furthermore, the feasibility of obtaining tissue-specific expression of an AAV-transduced gene has not been adequately addressed. Here, we provide experimental evidence to suggest that the vector assembly requires a precise signaling mechanism and that AAV infection of human cells is receptor-mediated. We also document erythroid lineage restricted expression following AAV-B19 hybrid vector-mediated transduction of primary human hematopoietic progenitor cells. Elucidation of the molecular details of these aspects of AAV biology will have important implications in the potential use of AAV as a vector in human gene therapy.

2 Encapsidation of Viral Genomes into Adeno-associated Virus Vectors Requires a Packaging Signal

Two AAV sequences are required for viral DNA replication. The first is the viral *rep* gene, which codes for four nonstructural proteins that are synthesized from a single open reading frame by the use of alternate promoters and splicing (Srivastava et al. 1983). The second is the viral origin of DNA replication which consists of a 145 nucleotide inverted terminal repeat (ITR) sequence (Lusby et al. 1980). Two of the viral Rep proteins (Rep78 and Rep68) are site-specific and strand-specific endonucleases that specifically bind to and cleave at the terminal resolution site (trs) within the AAV ITRs (Ashktorab and Srivastava 1989; Im and Muzyczka 1989, 1990, 1992; Snyder et al. 1990). AAV genomes with mutations in the *rep* gene are defective for viral DNA replication (Hermonat et al. 1984; Tratschin et al. 1984; Owens and Carter 1992). The terminal 125 nucleotides form a palindrome that can fold back on itself to form a T-shaped hairpin (HP) structure. The terminal HP is used as a primer for initiation of viral DNA replication (LUSBY et al. 1980; BERNS and BOHENZKY 1987; SRIVASTAVA 1987; MUZYCZKA 1992). Previous in vivo and in vitro studies have demonstrated that the intact ITRs are required in *cis* for AAV DNA replication as well as for rescue or excision from prokaryotic plasmids (SAMULSKI et al. 1982, 1983; SENAPATHY et al. 1984; GOTTLIEB and MUZYCZKA 1988; HONG et al. 1992).

Thus, it is clear that the ITRs are required for AAV DNA replication. However, the role of the individual AAV ITR structure in the viral genome rescue and/or packaging steps is less clear. In addition, the terminal 125 nucleotides within the ITR sequence are used to form the HP structure that serves as a primer of AAV DNA replication. However, the AAV ITRs also contain an additional 20 nucleotide stretch, designated as the D-sequence, that is not involved in HP formation. What role, if any, the D-sequence plays in AAV DNA rescue, replication, and packaging remains largely unexplored. To address this question, we constructed a number of recombinant AAV genomes containing deletions and substitutions in the D-sequence and studied in detail the effects of such alterations on AAV DNA replication, rescue, and packaging.

2.1 Rescue, Replication and Packaging of Adeno-associated Virus Genomes Containing D-sequence Deletions

The wild-type as well as the recombinant AAV genomes in which the D-sequences were either deleted or replaced with substitute sequences are depicted schematically in Fig. 1. The ITR consists of a terminal 125 nucleotides that form the HP structure which also contains the trs site at nucleotide 124 (NI et al. 1994) and the 20 nucleotide-long D-sequence (LUSBY et al. 1980). In plasmid pXS-18, both D-sequences as well as the two trs sites were deleted. These plasmids were transfected separately into human adenovirus 2 (Ad2)-infected KB cells and low M_r DNA samples were isolated at various times posttransfection. Equivalent amounts were digested with *Dpn*I to degrade unreplicated input plasmid DNA and analyzed on Southern blots using a ^{32}P-labeled probe specific for AAV DNA. These results are shown in Fig. 2. It is evident that rescue and replication of the AAV genome from plasmid pXS-18 was significantly reduced compared with that from plasmid p*Sub*201. We reasoned that the low efficiency of rescue/replication of the AAV genome from plasmid pXS-18 might be explained on the basis of loss of the trs site within the HP structure since *Bal*I restriction enzyme was used that deleted four nucleotides containing this site to generate this plasmid. Previous studies have documented the importance of this sequence in the resolution and replication of the AAV genome (IM and MUZYCZKA 1990; SNYDER et al. 1990, 1993; NI et al. 1994). The rationale for these experiments was that if the D-sequences per se, which are present as inverted repeat sequences within the two ITRs at each end, are indeed responsible for the efficient rescue/replication

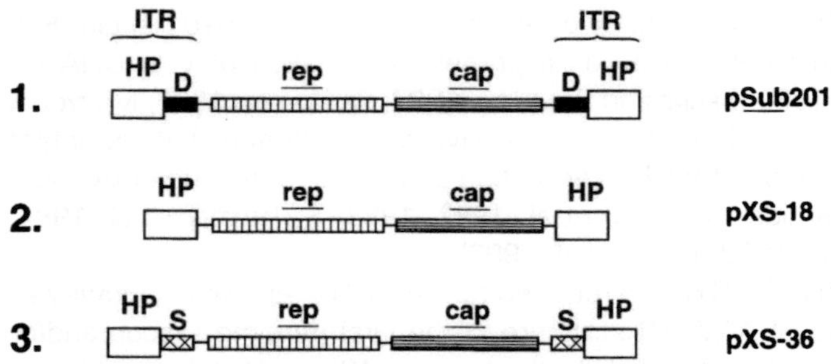

Fig. 1. Recombinant adeno-associated virus (AAV) genomes containing deletions/substitutions in the D-sequence within the viral inverted terminal repeats (*ITRs*). The palindromic hairpin (*HP*) structures and the D-sequences are illustrated as *open* and *closed boxes*, respectively. Plasmid p*Sub*201 has been described before (SAMULSKI et al. 1987), and the details of construction of the recombinant AAV plasmids pXS-18 and pXS-36 have been detailed elsewhere (WANG et al. 1995a, 1995b). The *cross-hatched box* represents a synthetic 20 nucleotide substitute sequence (S-sequence) that contains the terminal resolution site

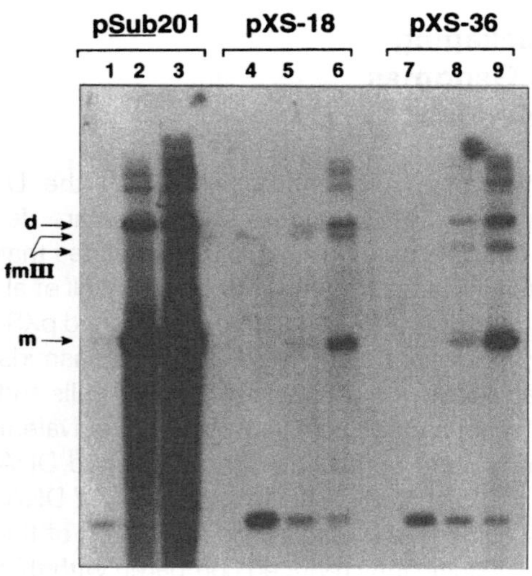

Fig. 2. Southern blot analysis of rescue and replication from plasmids p*Sub*201, pXS-18, and pXS-36. Low M_r DNA samples isolated by the method described by HIRT (1967), at 24 h (*lanes 1, 4, 7*), 48 h (*lanes 2, 5, 8*), and 72 h (*lanes 3, 6, 9*) posttransfection were digested with *Dpn*I, electrophoresed on agarose gels, and analyzed on Southern blots using the adeno-associated virus (AAV) DNA probe; *d* and *m* denote the dimeric and monomeric forms of the AAV DNA replicative intermediates, respectively, and *fmIII* represents the *Dpn* I-resistant, full-length plasmid DNA. Equivalent amounts of plasmids p*Sub*201 (*lanes 1–3*), pXS-18 (*lanes 4–6*), and pXS-36 (*lanes 7–9*) were analyzed. Deletion of both the D-sequences significantly reduces the efficiency of rescue/replication of the AAV genomes from plasmid pXS-18. Restoration of both the terminal resolution site (trs) by the S-sequences in plasmid pXS-36 is insufficient to compensate for the D-sequences without which approximately the same low level of rescue/replication of the AAV sequence occurs

of the AAV genome, this feature could be readily built back into the recombinant AAV genome. The two trs sites were restored in plasmid pXS-36 by inserting a synthetic oligonucleotide substitute (S)-sequence that contained the four nucleotides that were missing from the HP structure in pXS-18. This oligonucleotide also contained a non-AAV 20 nucleotide inverted repeat sequence that mimicked the D-sequence. It is interesting to note that the efficiency of the AAV rescue and replication from plasmid pXS-36 was not significantly enhanced compared with that from plasmid pXS-18 despite the presence of the two trs sequences and the two repeat sequences. In addition, rescue of the AAV genomes from plasmids pXS-18 and pXS-36 was not complete compared with that from plasmid p*Sub*201, since replication of full-length plasmid DNA (fmIII) was also evident. These results indicate that the D-sequence plays a crucial role in the efficient and precise rescue followed by selective replication of the AAV genome (WANG et al. 1995a)

Since the AAV genome could be rescued from plasmids pXS-18 and pXS-36, we wished to examine whether following DNA replication these aberrant genomes were also packaged into mature AAV virions. Equivalent volumes of culture supernatants from Ad2-infected KB cells transfected with these recombinant plasmids were digested exhaustively with DNase I to degrade any unencapsidated DNA, deproteinized to release the AAV DNA, and analyzed on quantitative DNA slot-blots using a ^{32}P-labeled AAV DNA probe. Such a blot is presented in Fig. 3A. A strong hybridization signal with virions produced from plasmid p*Sub*201 was detected, as expected. Surprisingly, however, no hybridization signal was obtained with plasmids pXS-18, and pXS-36, from which the AAV genome could be rescued. Prolonged exposures of these blots also failed to yield positive hybridization signals (data not shown). An additional possibility, that the efficiency of packaging was below the detection limit for AAV genomes rescued from these plasmids, was ruled out by the negative results obtained with the analyses of a 100-fold excess of supernatants from these cultures. Since the efficiency of rescue of the AAV genome from plasmid p*Sub*201 was not greater than 100-fold that from plasmids pXS-18 or pXS-36 (see Fig. 2), we conclude from these results that the D-sequence plays an important role in the successful packaging of the viral genomes into the progeny AAV virions (WANG et al. 1995b).

We next examined whether, following encapsidation, the AAV virions were infectious. Appropriate volumes of the culture supernatants collected and processed as described above were used to infect KB cells in the presence of Ad2. Equivalent amounts of low M_r DNA isolated 48 h post-infection were analyzed on Southern blots using a ^{32}P-labeled AAV DNA probe. It is apparent that the virions produced from plasmid p*Sub*201 were infectious since the characteristic viral DNA replicative intermediates could be readily observed (Fig. 3B). No replication occurred in the absence of coinfection with Ad2 (data not shown). Interestingly, no infectious virions were produced from plasmids pXS-18 and pXS-36, although the AAV genome could be rescued from these plasmids suggesting, once again, that in the absence of the D-sequences, these AAV genomes

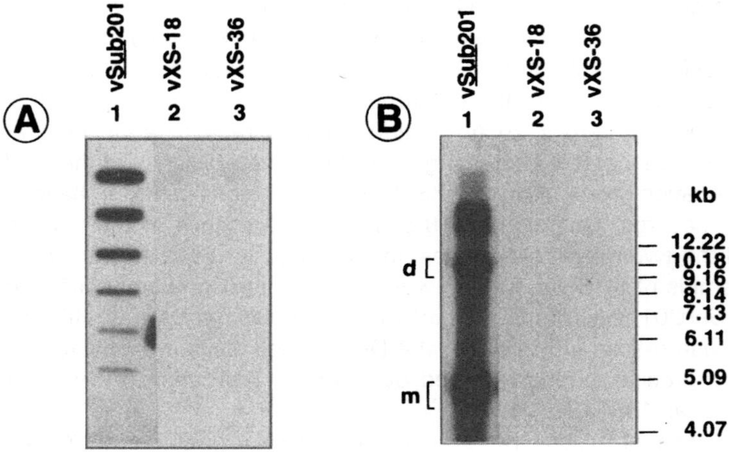

Fig. 3A,B. Packaging and biological activity of the progeny adeno-associated virus (AAV). **A** DNA slot-blot analysis of packaging of the AAV genomes was carried out with twofold dilutions of equivalent volumes of supernatants from cells transfected with the recombinant plasmids using the AAV DNA probe. Successful encapsidation of the AAV genomes rescued from plasmids p*Sub*201 (*lane 1*), but not from plasmids pXS-18 (*lane 2*), and pXS-36 (*lane 3*), occurs, suggesting that the D-sequence is crucial for packaging of the viral genomes. **B** Southern blot analysis of the efficiency of replication of the progeny AAV virions was carried out with culture supernatants containing the progeny virions generated from plasmids p*Sub*201 (*lane 1*), and approximately 50-fold larger volumes of that from plasmids pXS-18 (*lane 2*), and pXS-36 (*lane 3*) for secondary infections in the presence of Ad2. Equivalent amounts of the low M_r DNA samples isolated 72 h postinfection were analyzed on Southern blots using the AAV DNA probe. No AAV DNA replicative intermediates are seen with the virions generated from plasmids pXS-18 and pXS-36, presumably because no packaging of the viral genome occurs in the absence of the D-sequence

failed to be encapsidated. The lack of encapsidation of the AAV genomes from plasmids pXS-18 and pXS-36 could also be due to inadequate production of the viral capsid (*cap*) proteins which, in turn, would drastically reduce the packaging efficiency. To test this possibility, a recombinant AAV plasmid containing the CMV promoter-driven β-galactosidase gene (CMVp-*lac*Z) inserted between the two AAV-ITRs was cotransfected into Ad2-infected human 293 cells, and recombinant AAV-CMVp-*lac*Z virions were generated as previously described (Nahreini et al. 1993). The virus stocks were analyzed on quantitative DNA slot-blots using either *lac*Z or AAV DNA probes. Similar intensities of hybridization signals were obtained with the *lac*Z probe, but not with the AAV probe, indicating roughly equivalent production of the wild-type AAV-free, recombinant AAV-CMVp-*lac*Z virions, and approximately similar levels of expression of the viral *cap* genes from plasmids pAAV/Ad, pXS-18 and pXS-36 (data not shown). The biological activity of the recombinant AAV-CMVp-*lac*Z virions thus produced was evaluated as follows. Equivalent numbers of 293 cells were either mock-infected or infected at a multiplicity of infection (MOI) of 10 and stained with X-gal 48 h postinfection. Whereas no expression of the *lac*Z gene was apparent in mock-infected cells, abundant expression of the transduced gene was observed with the recombinant AAV virions generated either with pAAV/Ad (SAMULSKI et al. 1989), or with pXS-36 as the helper plasmid (WANG et al. 1995a), indicating that lack of expression of the *cap* genes from the latter plasmid could not account for the failure to encapsidate the recombinant AAV genome rescued from this plasmid. These results further strengthen our contention that the D-sequence may be the packaging signal for AAV.

How might the D-sequence play a role in packaging of the AAV genomes? Previous studies have documented that the AAV Rep proteins specifically interact with the AAV ITRs that are present in a cruciform structure (ASHKTORAB and SRIVASTAVA 1989; IM and MUZYCZKA 1989) and catalyze replication and resolution of the viral genome (IM and MUZYCZKA 1990, 1992). We reasoned that the D-sequence might also be involved in interacting with a viral protein(s) in order to facilitate efficient rescue, replication, and packaging of the AAV genomes. In this pursuit, whole cell extracts (WCEs) were prepared from uninfected, Ad2-infected, and Ad2+AAV coinfected HeLa cells, incubated with a ^{32}P-labeled D-sequence synthetic oligonucleotide probe and analyzed by electrophoretic mobility shift assays as previously described (ASHKTORAB and SRIVASTAVA 1989). These results are presented in Fig. 4. It is evident that the D-sequence formed a complex with a protein(s) present in uninfected cells and that this complex formation was specific, since the binding could be effectively competed by in-

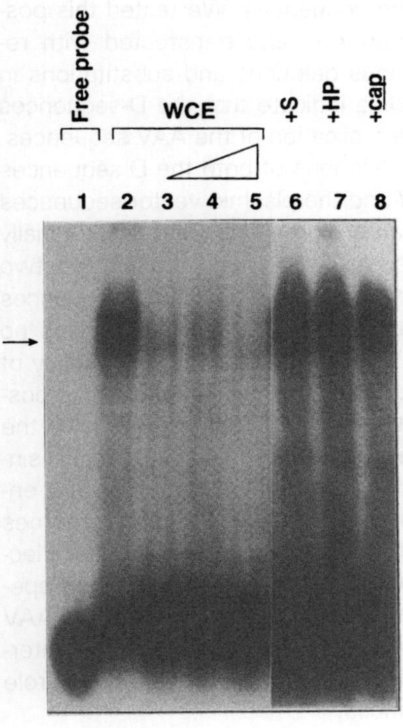

Fig. 4. Electrophoretic mobility shift assays for the D-sequence-binding protein. End-labeled-D-sequence probe (*lane 1*) was incubated with whole cell extracts (*WCEs*) prepared from mock-infected (*lane 2*) cells as well as in the presence of 25-fold (*lane 3*), 50-fold (*lane 4*), and 100-fold (*lane 5*) excess of unlabeled D-sequence oligonucleotide. The *arrow* indicates the formation of a specific DNA-protein complex. Competition experiments were also carried out in the presence of 100-fold excess of the S-sequence (*lane 6*), adeno-associated virus-human parvovirus (AAV-HP) sequence (*lane 7*), and AAV-*cap* sequence (*lane 8*) oligonucleotides. The D-sequence-protein complex formation (*lane 2*) is specifically competed with increasing concentrations of unlabeled D-sequence (*lanes 3–5*), but not with any of the other three sequences

creasing amounts of the unlabeled D-sequence probe. Similar results were obtained with WCEs prepared from Ad2-infected, and AAV+Ad2 coinfected cells (Wang et al. 1995b). Furthermore, this complex formation was not affected by incubations with a 100-fold excess of unlabeled DNA probes specific for the S-sequence, the AAV-HP sequence, or the AAV-*cap* sequences. These results document the existence of a hitherto unknown cellular protein(s) that specifically interacts with the D-sequence and which may be recruited by AAV to ensure efficient rescue, replication, and encapsidation of the viral genomes. We designate this protein as a D-sequence-binding protein (D-BP).

2.2 A Model for Selective Replication and Packaging of the Adeno-associated Virus Genomes

Previous studies have documented that, following excision, the cruciform structures of the ITRs are cross-linked to both ends of the linear duplex molecules of both the AAV and the plasmid vector sequences (Ward and Berns 1991; Hong et al. 1992), but only the AAV DNA is replicated. How might this be accomplished? Using an in vitro system, Ward and Berns (1991) hypothesized, but did not directly test, whether the inability of the plasmid sequences to undergo DNA replication was a consequence of the lack of the D-sequences from the AAV ITRs that were cross-linked to these sequences. We tested this possibility in intact cells that were infected with Ad2 and transfected with recombinant AAV plasmids that contained various deletions and substitutions in the D-sequence within the AAV ITRs. Our data indicate that the D-sequences indeed play an important role in the selective replication of the AAV sequences. This was evident by the fact that, following deletions of both the D-sequences from a wild-type AAV plasmid, both the AAV and the plasmid vector sequences underwent DNA replication nearly to the same extent following what initially appeared to be less than optimal rescue presumably due to deletions of the two trs sites. However, when the two trs sites were restored and the D-sequences substituted by a non-AAV synthetic oligonucleotide sequence, there was no appreciable increase in the extent of the AAV DNA replication, the efficiency of which was nearly identical to that of the vector sequences. An alternative possibility, that the observed lack of encapsidation is due to failure to resolve the replicative intermediates generated from plasmids pXS-18 and pXS-36 into single-stranded viral genomes, is less likely in view of the fact that abundant encapsidation was observed despite the apparent absence of these genomes generated from additional recombinant plasmids(Wang et al. 1995a). The electrophoretic mobility shift assays revealed that the D-sequence interacted specifically with a protein(s) present in uninfected cells. Therefore, neither the AAV Rep proteins nor the Ad2-encoded gene products were involved in this interaction. Taken together, our data suggest that the D-sequence plays a critical role in the selective replication of the AAV sequences.

We propose a model (Fig. 5) for AAV rescue, replication and packaging that appears to encompass all available data. For the sake of clarity, rescue and replication of only the left AAV ITR (HP+D-sequence, or HP+S-sequence) cross-linked to the vector sequences in plasmids p*Sub*201 (Fig. 5, step a) and pXS-36 (Fig. 5, step f), respectively, are shown. First, the HP forms a Holliday structure (Holliday 1964) which is cleaved symmetrically by the AAV Rep proteins (steps b, g) leading to excision of the AAV and the vector sequences that are cross-linked to the cruciform structures of the AAV ITRs (steps c, h), except that only the AAV sequences retain the D-sequence (from p*Sub*201) and the S-sequence (from pXS-36).The vector DNA lacks these sequences. A D-BP that specifically interacts with the D-sequence, but not with the S-sequence, catalyzes the AAV DNA replication in a preferential manner (step d). In the absence of the interaction with the D-BP, both the AAV and the vector sequences undergo DNA replication, but the efficiency of replication is significantly reduced (step i). And finally, the D-sequence is utilized as a packaging signal to selectively encapsidate the single-stranded viral genomes into the progeny AAV virions (step e), and the absence of this putative packaging signal results in the failure to encapsidate

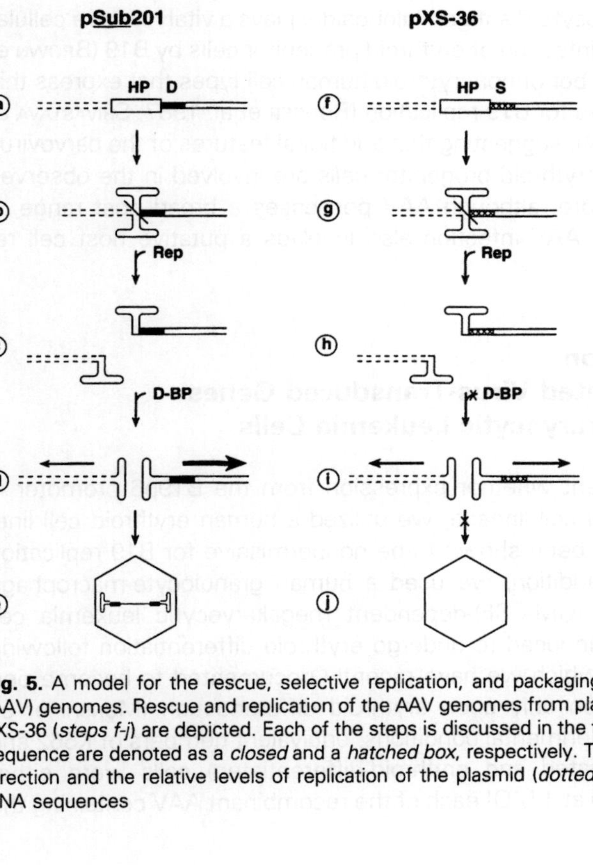

Fig. 5. A model for the rescue, selective replication, and packaging of the adeno-associated virus (AAV) genomes. Rescue and replication of the AAV genomes from plasmids p*Sub*201 (*steps a-e*), and pXS-36 (*steps f-j*) are depicted. Each of the steps is discussed in the text. The D-sequence and the S-sequence are denoted as a *closed* and a *hatched box*, respectively. The *horizontal arrows* reflect the direction and the relative levels of replication of the plasmid (*dotted lines*) and the AAV (*solid lines*) DNA sequences

(step j). The understanding of the precise molecular mechanism of viral assembly must await further experimental analyses.

3 Transduction of Human Hematopoietic Cells by Adeno-associated Virus Vectors Is Receptor-Mediated

As indicated above, the human parvovirus B19 exhibits a remarkable tissue tropism for the erythroid lineage in human hematopoietic progenitor cells. So far, only primary cells from human bone marrow (Ozawa et al. 1986, 1987; Srivastava and Lu 1988; Srivastava et al. 1989, 1990; Takahashi et al. 1990), fetal liver (Yaegashi et al. 1989), peripheral blood (Schwarz et al. 1992), and umbilical cord blood (Srivastava et al. 1992) have been shown to support B19 replication in vitro. Recently, we (Munshi et al. 1993) and others (Shimomura et al. 1992) have identified megakaryocytic leukemia cell lines which become permissive for B19 replication upon treatment with erythropoietin (Epo). Recently, it has been suggested that the erythrocyte P antigen (globoside) plays a vital role as a cellular receptor for the selective infection of erythroid progenitor cells by B19 (Brown et al. 1993). However, a number of nonerythroid human cell types that express this receptor are nonpermissive for B19 replication (Rouger et al. 1987; Srivastava et al. 1990; Brown et al. 1994), suggesting that additional features of the parvovirus B19 and primary human erythroid progenitor cells are involved in the observed tissue tropism. Furthermore, although AAV possesses a broad host range, it remains unclear whether AAV infection also involves a putative host cell receptor.

3.1 Lack of Expression of Adeno-associated Virus-Transduced Genes in Human Megakaryocytic Leukemia Cells

In our studies to document whether expression from the B19p6 promoter is restricted to the erythroid cell lineage, we utilized a human erythroid cell line, K562, that has previously been shown to be nonpermissive for B19 replication (Ozawa et al. 1986). In addition, we used a human granulocyte-macrophage colony-stimulating factor (GM-CSF)-dependent megakaryocytic leukemia cell line, MB-02, that can be induced to undergo erythroid differentiation following treatment with Epo and which we have recently documented to become permissive for B19 replication only upon erythroid differentiation (Munshi et al. 1993). Under identical experimental conditions, equivalent numbers of K562 and MB-02 (both undifferentiated and erythroid-differentiated) cells were either mock-infected, or infected at 1 MOI each of the recombinant AAV containing the firefly luciferase (Luc) reporter gene driven by the herpesvirus thymidine kinase (TK) and the B19p6 promoters. Cells were infected with vTKp-Luc or vB19p6-Luc virions separately and analyzed for Luc gene expression (Fig. 6). It is evident that whereas significant expression of the Luc gene occurred from the TK promoter in K562 cells, no significant activity from the B19p6 promoter could be detected (panel A). It is intriguing to note that no expression from either of these promoters occurred in either undifferentiated (panel B), or erythroid-differentiated (panel C) MB-02 cells. Although we initially interpreted these results to mean that these promoters were nonfunctional in MB-02 cells, our subsequent studies described below revealed that these cells were nonpermissive for AAV infection presumably because they lack the putative cellular receptors for AAV.

We also examined an additional human megakaryocytic leukemia cell line, M07e, for susceptibility of infection by the recombinant AAV virions. This was carried out by using two different recombinant AAV viral stocks containing the Luc and *lacZ* genes under the control of the cytomegalovirus (CMV) promoter.

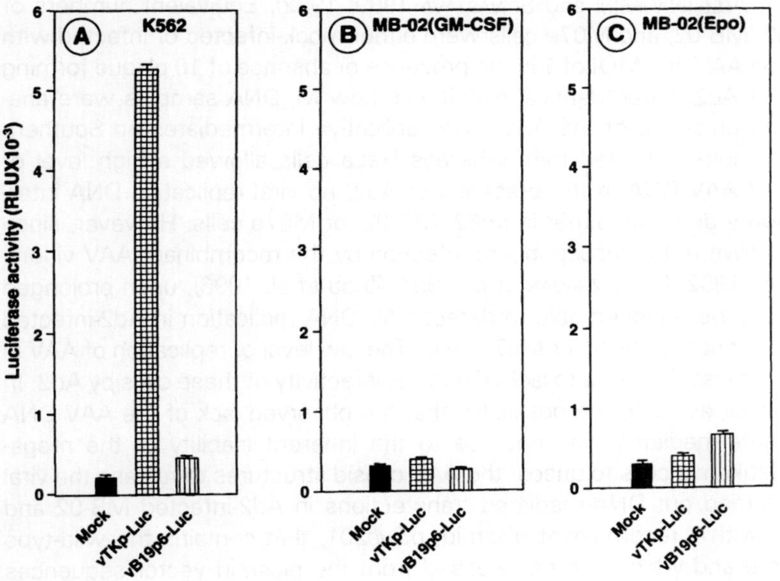

Fig. 6A–C. Comparative analysis of expression of luciferase (Luc) activity following recombinant adeno-associated virus (AAV)-mediated transduction of K562 and MB-02 cells. **A** K562 cells that were either mock-infected or infected with the indicated recombinant AAV virions were analyzed for Luc activity. **B** Undifferentiated MB-02 cells grown in the presence of the human granulocyte-macrophage colony-stimulating factor (GM-CSF) were similarly transduced and analyzed. **C** MB-02 cells were induced to undergo erythroid differentiation following treatment with erythropoietin Epo, transduced with the recombinant AAV virions and analyzed as above. Little expression from the B19p6 promoter occurs following AAV-mediated transduction of K562 cells, known to be permissive for AAV infection but non-permissive for B19 replication. Erythroid-differentiated MB-02 cells, known to be permissive for B19 replication, also do not allow expression from this promoter

Prior to these analyses, a high level expression from the CMV promoter in these cells was documented by plasmid DNA transfections to ensure the functional activity of this promoter (data not shown). Equivalent numbers of K562, MB-02, and M07e cells were either mock-infected, or infected separately with vCMVp-Luc or vCMVp-*lacZ* virions at an MOI of 1 and analyzed for the expression of the reporter genes. The results indicated that whereas a significantly high level of Luc activity was detected in the recombinant virus-transduced K562 cells, no detectable activity was obtained in MB-02 or M07e cells. Similarly, whereas *lacZ* gene expression in a number of the recombinant virus transduced K562 cells was readily observed, no such activity could be detected in the transduced MB-02 or M07e cells (PONNAZHAGAN et al.1996). These results further substantiate our contention that these megakaryocytic leukemia cells are resistant to infection by AAV because they lack the functional receptors for AAV.

Since it remained possible that expression of the transduced genes was below the detection limit, we next wished to examine whether the wild-type AAV could undergo replication in these cells in the presence of a helper virus, such as Ad2, since previous studies have established abundant replication and accumulation of the viral DNA in permissive human cells, such as HeLa, in the presence of Ad2 (NAHREINI and SRIVASTAVA 1989, 1992). Equivalent numbers of HeLa, K562, MB-02, and M07e cells were either mock-infected or infected with the wild-type AAV at a MOI of 1 in the presence or absence of 10 plaque forming units (pfu) of Ad2 under identical conditions. Low M_r DNA samples were analyzed for the presence of the AAV DNA replicative intermediates on Southern blots. The results indicated that, whereas HeLa cells allowed a high level of replication of AAV DNA in the presence of Ad2, no viral replicative DNA intermediates were detected either in K562, MB-02, or M07e cells. However, since K562 are known to be susceptible to infection by the recombinant AAV virions (WALSH et al. 1992; PONNAZHAGAN et al. 1994; ZHOU et al. 1996), upon prolonged exposure, we were indeed able to detect AAV DNA replication in Ad2-infected K562 cells, but not in MB-02 or M07e cells. The low level of replication of AAV in K562 cells is most likely due to lack of optimal infectivity of these cells by Ad2. In order to further evaluate the possibility that the observed lack of the AAV DNA replicative intermediates was not due to the inherent inability of the megakaryocytic leukemia cells to uncoat the AAV capsid structures to release the viral DNA, we carried out DNA-mediated transfections in Ad2-infected MB-02 and M07e cells with a recombinant plasmid, p*Sub*201, that contains the wild-type AAV genome and which can be released from the plasmid vector sequences following digestion with the restriction endonuclease *Pvu*II (SAMULSKI et al. 1987). HeLa and K562 cells were also analyzed under identical conditions. The results of these experiments indicated abundant replication of the transfected AAV DNA in HeLa cells consistent with previous studies (NAHREINI and SRIVASTAVA 1989, 1992). Upon prolonged exposure, the viral DNA replicative intermediates could also be visualized in K562 cells. No replication of AAV DNA occurred in MB-02 and M07e cells although these cells are permissive for Ad2 infection (data not shown).

3.2 Expression from the B19p6 Promoter in Cells Undergoing Erythroid Differentiation

Since we were unable to transduce MB-02 cells with the recombinant AAV virions, we next wished to examine whether expression from the B19p6 promoter correlated with the differentiated state of these cells especially since erythroid-differentiated MB-02 cells support B19 gene expression and replication (MUNSHI et al. 1993). Transfections were carried out with undifferentiated and erythroid differentiated MB-02 cells with the B19p6-Luc plasmid. In addition, a plasmid containing an erythroid-specific enhancer element, HS-2, the DNase I hypersensitive site 2 of the locus control region (LCR) of the human β-globin gene cluster (TUAN et al. 1989), was used to evaluate whether this enhancer could modulate expression from the B19p6 promoter in an erythroid cell-specific manner. K562 cells were also included in these experiments. These results are presented in Fig. 7. It is evident that expression from the B19p6 promoter was significantly increased by the HS-2 enhancer element in K562 cells (panel A). Interestingly, little expression from the B19p6 promoter, either with or without the HS-2 enhancer, could be detected in undifferentiated MB-02 cells (panel B).

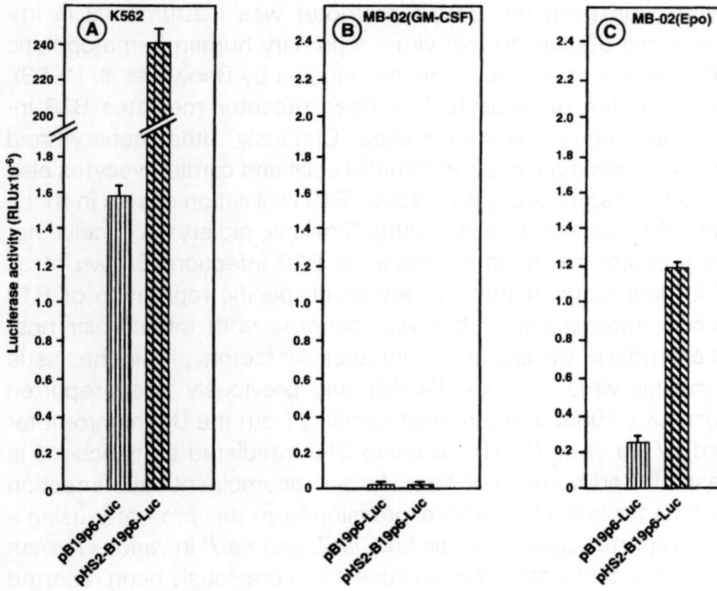

Fig. 7A–C. Comparative analysis of expression from the B19 promoters in K562 and MB-02 cells following DNA-mediated transfections. **A** K562 cells were transfected separately with the indicated recombinant adeno-associated virus (AAV) plasmids and analyzed for luciferase (Luc) activity. Undifferentiated MB-02 cells **B**, or erythroid-differentiated MB-02 cells **C** were also analyzed under identical conditions. Abundant expression from the B19p6 promoter occurs in K562 cells which is further augmented by the HS-2 enhancer following DNA-mediated transfection. Expression from this promoter also occurs in MB-02 cells but only upon erythroid differentiation

However, following erythroid differentiation of these cells, significant expression from this promoter was detected, the extent of which was further augmented by the HS-2 enhancer in these cells (panel C). We noted little difference in expression from the B19p6 promoter following DNA-mediated transfections with plasmids that either lacked or contained the AAV ITRs (data not shown).

We also wished to examine whether the lack of expression from the B19p6 promoter in undifferentiated MB-02 cells was due to the presence of a nonspecific inhibitor of Luc activity. Two additional recombinant plasmids containing the Luc gene under the control of the CMV and the SV40 promoters were used in transfection experiments under identical conditions. The results of these experiments indicated that abundant expression from these promoters occurred in K562 cells as previously observed. Significant expression from the CMV promoter as well as the SV40 promoter, but not from the B19p6 promoter, with or without the upstream HS-2 enhancer element, also occurred in MB-02 cells. Taken together, these results indicate that the B19p6 promoter function may be regulated by factors expressed in cells undergoing erythroid differentiation and that the observed lack of expression from the B19p6 promoter in undifferentiated cells is not due to the presence of a nonspecific inhibitor in MB-02 cells (Ponnazhagan et al. 1996).

Previous studies from our laboratory with a recombinant AAV-B19 hybrid indicated that expression from the B19p6 promoter was instrumental in imparting the erythroid cell tropism to this virus in primary human hematopoietic progenitor cells (Srivastava et al. 1989). Recent studies by Brown et al. (1993), suggest, however, that the erythrocyte P antigen receptor mediates B19 infection of cells in the human erythroid lineage. Curiously, other nonerythroid human cells such as megakaryocytes, endothelial cells and cardiomyocytes also express the P antigen receptor and yet no active B19 replication occurs in these cells (Rouger et al. 1987; Srivastava et al. 1990). Similarly, pig erythroid cells that also express this receptor are nonpermissive for B19 infection (Brown et al. 1994). All available data suggest that the erythroid-specific replication of B19 results from specific interactions of the viral genome with intracellular host factors. A cogent example of the crucial role intracellular factors play in the tissue tropism of the minute virus of mice (MVM) has previously been reported (Spalholz and Tattersall 1983). Indiscriminate activity from the B19p6 promoter has been reported by Liu et al. (1991) following DNA-mediated transfections in nonpermissive, established human cell lines. In our recombinant virus infection studies, we have thus far failed to obtain expression from this promoter using a number of different reporter genes such as Luc, *lacZ*, and *neo*^{R} in various human cell lines. This is an important distinction because it has previously been reported that, following DNA-mediated transfections, on the order of 500 000 DNA molecules/cell are required to accomplish detectable expression of the transfected genes (Curiel et al. 1992). Thus, in marked contrast to virus infections, DNA-mediated transfections may not be physiologically relevant, and all gene expression data obtained by the latter means must be interpreted with caution (Liu et al. 1991, 1992).

The aforementioned problems notwithstanding, we were still able to obtain erythroid-specific expression from the B19p6 promoter in MB-02 cells. MB-02 cells offered a unique model system primarily because the same culture system could be examined under both undifferentiated and terminally differentiated states, and we have previously reported that the erythroid-differentiated MB-02 cells become permissive for expression of the B19 genes and that complete replication of the viral genome occurs (MUNSHI et al. 1993). Although expression from the B19p6 promoter could be detected in K562 cells, the significance of this observation is less clear because the wild-type B19 does not infect these cells, and little expression from this promoter occurs in the context of a recombinant AAV genome. Expression from the B19p6 promoter only in erythroid-differentiated MB-02 cells, however, may be important since it correlates well with the replication of the B19 viral genome in these cells (MUNSHI et al. 1993). It is important to emphasize that in these experiments only a small percent of the MB-02 cells underwent successful erythroid differentiation, as determined by the morphologic criteria, which could partially account for the observed low levels of expression from the B19p6 promoter compared to that in K562 cells. Thus, it is tempting to speculate that the intracellular factors present in erythroid-differentiated MB-02 cells contribute to the observed tropism of the parvovirus B19. The fact that expression from the B19p6 promoter was significantly augmented by the erythroid-specific HS-2 enhancer, but only in erythroid-differentiated MB-02 cells, further strengthens this hypothesis. Clearly, the potential role of other ubiquitous *cis*-acting elements in the activation of the B19p6 promoter in the nonpermissive K562 cells remains to be examined.

Taking into account all available data, we now propose that the remarkable erythroid cell tropism of the parvovirus B19 requires a dual level of specificity. First, the erythrocyte P antigen is utilized as a cellular receptor for viral entry effectively into erythroid cells and, to some extent, into certain nonerythroid cells. Second, optimal transcriptional *trans*-activation of the B19p6 promoter is accomplished by intracellular factors that are highly restricted to human erythroid cells. The proposed mechanism thus provides an explanation for the failure of B19 to undergo productive replication in nonerythroid cells that express the P antigen receptor and, therefore, permit viral entry. However, no viral gene expression occurs since these nonerythroid cells lack the specific intracellular factors required for optimal transcriptional *trans* activation of the B19p6 promoter. This proposal also helps explain the retention of the erythroid tropism of a recombinant AAV-B19 hybrid virus that bypasses the P antigen receptor, and yet undergoes productive replication in an erythroid cell-specific manner, because of optimal transcriptional *trans* activation of the B19p6 promoter only in erythroid cells (SRIVASTAVA et al. 1989).

In our attempts to transduce MB-02 cells with the recombinant AAV vectors, we discovered that these cells were nonpermissive for AAV-infection. Several lines of evidence supported this contention. No significant activity of any of the AAV-transduced genes such as the reporter genes Luc and *lac*Z was evident in these cells. No G418-resistant clones of MB-02 cells could be obtained following AAV-mediated transduction with a selectable gene such as *neo*^{R}. We also failed to detect the wild-type AAV DNA replicative intermediates even with coinfection with the wild-type Ad2. In addition, infections of MB-02 cells at both low and high MOI of the wild-type AAV followed by serial passage failed to reveal the presence of the integrated proviral AAV genomes on Southern blots. An additional human megakaryocytic leukemia cell line, M07e, yielded identical results (data not shown). Thus, to our knowledge, this is the first instance in which a human cell type has been identified that cannot be infected by AAV. Although the broad host range of AAV that transcends species barrier is well-documented (Muzyczka 1992), we now propose that AAV infection of human cells is mediated by a hitherto unknown cellular receptor.

4 The Adeno-associated Virus-B19 Hybrid Vector Allows Erythroid Cell-Specific Expression of Transduced Genes

Although the target cell specificity for B19 infection has been suggested to be mediated by the erythrocyte P antigen (Brown et al. 1993), a number of nonerythroid cells that express this receptor are nonpermissive for B19 replication (Brown et al. 1994; Rouger et al. 1987; Srivastava et al. 1990). In addition, whereas indiscriminate expression from the B19p6 promoter following DNA-mediated transfection nonerythroid established cell lines has been reported (Liu et al. 1991), in all our recombinant virus infection studies to date, we have failed to obtain expression from this promoter in nonerythroid cells (Ponnazhagan et al. 1996; Wang et al. 1995c; Zhou et al. 1996).

In order to resolve this apparent paradox, we examined expression from the B19p6 promoter following both DNA-mediated transfections and recombinant virus infections of a nonerythroid human nasopharyngeal carcinoma cell line, KB, that is nonpermissive for B19 replication. We also utilized a human megakaryocytic leukemia cell line, MB-02, that can be induced to undergo erythroid differentiation following treatment with Epo and that we have recently documented to become permissive for B19 replication (Munshi et al. 1993). MB-02 cells thus offered a useful system to evaluate the B19 promoter functions under both permissive and nonpermissive conditions. Here, we provide evidence that little expression from the B19p6 promoter occurs in the context of a recombinant AAV genome in nonerythroid cells known to be permissive for the viral infection.

4.1 Differential Expression from the B19p6 Promoter in Nonpermissive Human Cells

We have previously reported that expression from the B19p6 promoter is largely restricted to primary human hematopoietic cells in the erythroid lineage following infection with a recombinant AAV-B19 hybrid virus (Srivastava et al. 1989). However, as noted above, Liu et al. (1991) concluded from plasmid transfection studies that expression from the B19p6 promoter occurs indiscriminately in nonpermissive cells. In order to resolve this issue, we performed both DNA-mediated transfections with equivalent amounts of plasmids (3 μg pB19p6-Luc, in addition to a *lacZ* plasmid to control for transfection efficiency) and recombinant AAV infections (vB19p6-Luc at 1 moi) of nonerythroid human KB cells. Mock transfections/infections as well as transfections/infections with the TKp-Luc construct were also used as appropriate controls. These results are shown in Fig. 8. It is evident that the TK promoter-driven expression of the Luc gene was exceeded by that from the B19p6 promoter following DNA-mediated transfections (panel A). However, no Luc activity from the latter promoter could be detected following recombinant virus infection, although expression from the TK promoter was still detectable under identical conditions. Similar results were obtained at MOIs of 10 and 100 indicating that the observed lack of expression was not due to the limiting viral titers (data not shown). These results indicate that expression from the B19p6 promoter does not occur in nonerythroid cells in the context of a viral genome. The lack of expression from the B19p6 promoter has been substantiated in an additional nonerythroid cell line, HeLa, transduced with the recombinant AAV-B19p6-Luc virions and with two different re-

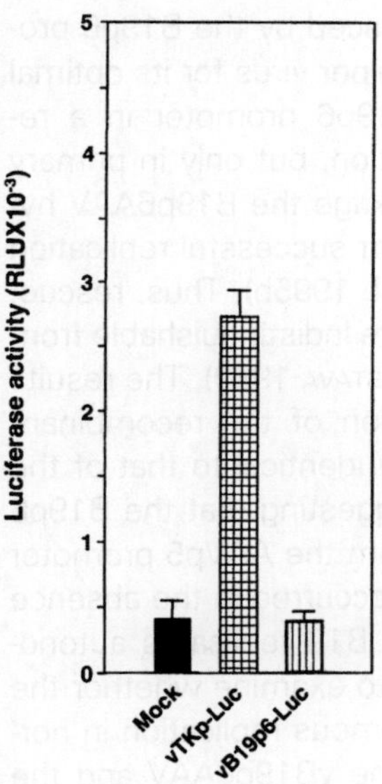

Fig. 8. Comparative analysis of expression of luciferase (Luc) activity from the TK and the B19p6 promoters in human KB cells. Human KB cells were either mock-infected or infected at multiplicity of infection of 1 with the indicated recombinant adeno-associated virus (AAV) virions. Luc activity was determined 48 h postinfection as described above. Expression from the B19p6 promoter does not occur in nonerythroid human cells following AAV-mediated transduction

combinant AAV virions containing the selectable marker gene for resistance to neomycin (*neo*^{R}) and the *lac*Z reporter gene under the control of this promoter (unpublished results). Thus, under nonselective conditions, little expression of the transduced genes occurs from the B19p6 promoter introduced into established human cell lines in the context of a recombinant parvoviral genome.

4.2 Erythroid Cell Specificity of the Hybrid Vector

Previous studies have established that two key components are required for successful replication of the human parvoviral genome. These include the viral ITR sequences that serve as the origin of DNA replication (LUSBY et al. 1980; SRIVASTAVA 1987) and the viral nonstructural (NS) proteins called Rep for AAV and NS-1 for B19 (SRIVASTAVA et al. 1983; SHADE et al. 1986). In AAV, expression of the major Rep proteins occurs from the p5 promoter (HERMONAT et al. 1984), whereas in B19 the NS-1 gene is expressed from the p6 promoter (OZAWA et al. 1987). Previous studies have also established that the cloned wild-type AAV genome undergoes DNA replication in human cells in the presence of a helper virus (SAMULSKI et al. 1982, 1983; NAHREINI and SRIVASTAVA 1989, 1992). However, the cloned B19 genome is not infectious because of deletions in its ITR sequences (SHADE et al. 1986). We have previously reported the construction of a recombinant AAV-B19 genome, in which the B19 ITRs were replaced by the AAV ITRs (SRIVASTAVA et al. 1989). This hybrid virus was biologically indistinguishable from the wild-type B19, demonstrating autonomous replication in and cytotoxicity to human erythroid progenitor cells (SRIVASTAVA et al. 1989).

In order to unequivocally establish the erythroid cell specificity of expression from the B19p6 promoter, we constructed a novel AAV genome in which only the authentic promoter at map unit 5 (AAVp5) was replaced by the B19p6 promoter (Fig. 9). Although the wild-type AAV requires a helper virus for its optimal replication, we hypothesized that insertion of the B19p6 promoter in a recombinant AAV would permit autonomous viral replication, but only in primary human erythroid progenitor cells. We were able to package the B19p6AAV hybrid genomes into mature AAV virions, which underwent successful replication in KB cells, but only in the presence of Ad2 (WANG et al. 1995b). Thus, rescue, replication and packaging of the B19p6AAV genome were indistinguishable from that of the wild-type AAV in KB cells (NAHREINI and SRIVASTAVA 1989). The results also indicated that the extent of rescue and replication of the recombinant B19p6AAV genome in the presence of Ad2 was nearly identical to that of the wild-type AAV genome from the plasmid p*Sub*201, suggesting that the B19p6 promoter was *trans*-activated by Ad2, similar to that from the AAVp5 promoter (SAMULSKI et al. 1982, 1983). No rescue and replication occurred in the absence of Ad2 infection (WANG et al. 1995b). Since parvovirus B19 replicates autonomously in human hematopoietic cells, we next wished to examine whether the recombinant B19p6AAV virions were capable of autonomous replication in normal human bone marrow cells. Equivalent stocks of the vB19p6AAV and the

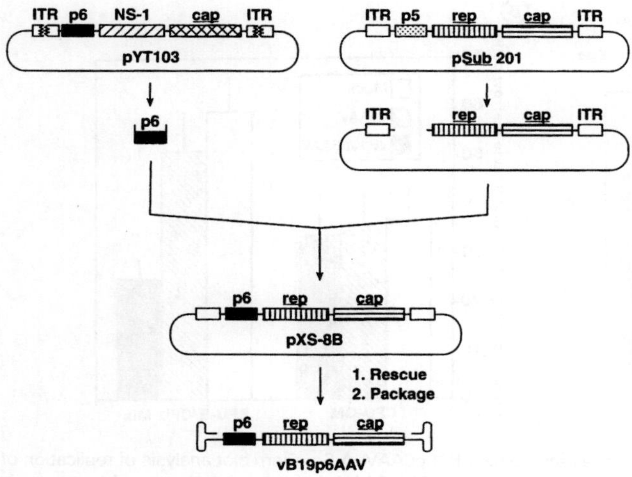

Fig. 9. Strategy for construction of the recombinant B19p6AAV plasmid and virions. Details of the construction of these plasmid are provided elsewhere (WANG et al. 1995b). Standard cloning techniques were used as described by SAMBROOK et al. (1989). The B19p6 promoter was derived from plasmid pYT103 (SHADE et al. 1986). A recombinant adeno-associated virus (AAV) plasmid, p*Sub*201, has previously been described (SAMULSKI et al. 1987). The B19p6 and the AAVp5 promoters are depicted as *closed* and *stippled* rectangles, respectively. The B19 inverted terminal repeats (ITRs) are shown as *broken rectangles* indicating deletions within these sequences (SHADE et al. 1986). The AAV-ITRs are depicted as *open rectangles*, and the AAV *rep* and *cap* gene sequences are represented by *boxes* with *vertical lines* and *horizontal lines*, respectively. The recombinant vB19p6AAV virions were generated from plasmid pXS-8B as previously described. (From SAMULSKI et al. 1989; NAHREINI et al. 1993)

wild-type AAV were used separately to infect low density human bone marrow (LDBM) cells in the presence and absence of Epo. It is interesting to note that, whereas no replication of the wild-type AAV was detected, either in the absence or the presence of Epo, the vB19p6AAV underwent successful, autonomous replication in LDBM cells, and the effect was quantitatively augmented in the presence of Epo (Fig. 10A). The autonomous replication of the recombinant B19p6AAV virions was remarkably similar to that of the wild-type parvovirus B19 (OZAWA et al. 1986, 1987; SRIVASTAVA et al. 1989, 1990). These results indicate that the B19p6 promoter sequence is necessary to confer autonomous replication competence to AAV in primary human hematopoietic progenitor cells.

The parvovirus B19 has previously been shown to selectively inhibit colony formation by erythroid progenitor cells in human bone marrow (OZAWA et al. 1986, 1987; SRIVASTAVA and LU 1988; SRIVASTAVA et al. 1989, 1990). AAV, by contrast, can infect myeloid as well as erythroid hematopoietic progenitors cells (ZHOU et al. 1993, 1994). Hematopoietic progenitor cell assays (colony-forming unit - granulocyte-macrophage, CFU-GM; burst-forming unit - erythroid, BFU-E; colony-forming unit - mixed cells, CFU-Mix) were carried out following mock infection, or infections with the wild-type AAV or the vB19p6AAV virions under identical conditions (MORITZ et al. 1994). Whereas CFU-GM colony formation was

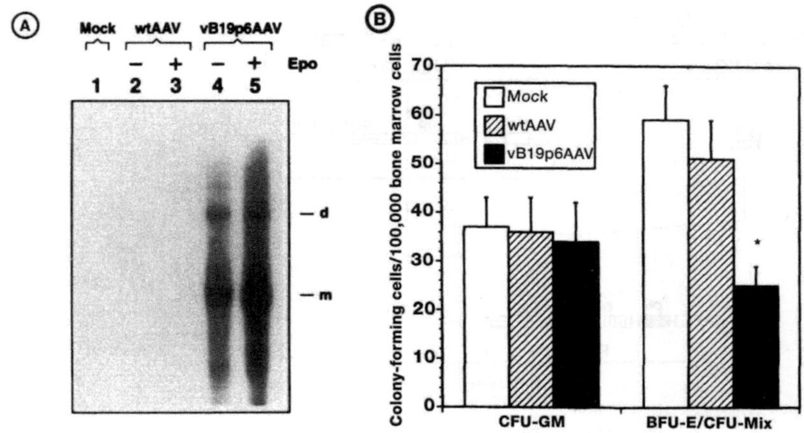

Fig. 10A,B. Biological activity of the recombinant B19p6AAV. **A** Southern blot analysis of replication of the wild-type adeno-associated virus (AAV) and the vB19p6AAV virions in normal human bone marrow cells was carried out with equivalent amounts of low M_rDNA samples isolated from mock-infected (*lane 1*), wild-type AAV-infected (*lanes 2 and 3*), or the vB19p6AAV-infected (*lanes 4 and 5*) cells grown in the absence or presence of erythropoietin (Epo); *m* and *d* denote the monomeric and dimeric forms of the replicative viral DNA intermediates, respectively. **B** Effect of the wild-type AAV and the vB19p6AAV virions on colony formation (colony-forming unit – granulocyte-macrophage, CFU-GM; burst-forming unit – erythroid, BFU-E, and colony-forming unit – mixed-cell lineage, CFU-Mix) by normal human bone marrow progenitor cells was carried out following mock infection (*open bars*) wild-type AAV infection (*hatched bars*), and vB19p6AAV infection (*closed bars*), as described by Moritz et al. (1994). These results represent data obtained from six different culture plates. Values indicated are the mean ± SD. $^*p < 0.001$, for vB19p6AAV vs the wild-type AAV and mock-infected cells. The wild-type AAV does not undergo DNA replication in primary human bone marrow cells and is not cytotoxic to hematopoietic progenitor cells (Srivastava et al. 1989, 1990). The recombinant B19p6AAV, by contrast, is capable of autonomous replication and is selectively cytotoxic to erythroid progenitor cells, indicating that the B19p6 promoter is necessary and sufficient to confer autonomous replication competence as well as erythroid cell specificity to AAV

not affected, infection with the vB19p6AAV virions significantly reduced BFU-E/CFU-Mix colony formation in hematopoietic progenitor cell assays (Fig. 10B). Again, the wild-type AAV had no effect on CFU-GM and BFU-E colony formation. These results indicate that the B19p6 promoter is sufficient to impart erythroid cell tropism to AAV in primary human hematopoietic progenitor cells. This simple substitution is necessary and sufficient to allow the otherwise noninfectious AAV to undergo successful, autonomous replication in the erythroid lineage in primary human hematopoietic progenitor cells. Thus, expression from the B19p6 promoter is crucial for establishing erythroid cell tropism.

It is becoming increasingly clear that the P antigen receptor alone may be insufficient to account for the permissive state of the target cell. Several lines of evidence support this notion. First, a number of nonerythroid human cell types that express this cell surface receptor are nonpermissive for B19 replication (Rouger et al. 1987; Srivastava et al. 1990). Second, the erythroid progenitor cells from pigs and monkeys that express the P antigen also do not support B19 replication (Brown et al. 1994). Third, a hybrid AAV-B19 genome that is packaged into AAV virions, and therefore bypasses the P antigen receptor, undergoes replication in bone marrow cells and retains its erythroid cell tropism (Srivastava et al. 1989). In addition, previous studies with the MVM have established that, although viral entry is mediated by different specific cell surface receptors, the target cell specificity of replication of MVM is dependent on the intracellular host cell factors (Spalholz and Tattersall 1983). Furthermore, in view of the fact that the P antigen is expressed in high quantities in mature erythrocytes (Marcus et al. 1981), it is difficult to envisage an efficient infection by B19 in vivo since these cells lack functional nuclei. Curiously, detectable levels of this receptor have not been observed on erythroid progenitors (BFU-E and CFU-E) and proerythroblasts (von dem Borne et al. 1986), and, since numerous studies have identified these to be the B19-targets (Brown et al. 1994; Ozawa et al. 1986, 1987; Srivastava and Lu 1988; Srivastava et al. 1989, 1990), it would also seem difficult to explain efficient B19 infection in vitro mediated by the P antigen.

These apparent inconsistencies notwithstanding, it is interesting to note that indiscriminate activity from the B19p6 promoter has previously been reported (Liu et al. 1991). It should be emphasized, however, that those studies were carried out by plasmid-mediated transfections in established human cell lines. In addition to the studies reported here, our cumulative data to date with recombinant AAV infections using a number of reporter genes, such as *neo*^{R}, *lacZ* and Luc, have revealed little expression from the B19p6 promoter in established human cell lines that are permissive for the recombinant viral infections (Ponnazhagan et al. 1996). However, expression from the B19p6 promoter in primary human hematopoietic cells has been detected (Wang et al. 1995c). Taken together, these studies further underscore the importance of exercising caution in interpreting the expression data obtained with plasmid transfections (Liu et al. 1991, 1992).

It is intriguing that, in previous studies by Labow and Berns (1988), an AAV hybrid genome, in which the AAVp5 promoter was replaced with the SV40 promoter, was shown to undergo rescue and replication in the presence of Ad2 in KB cells, but expression from the SV40 promoter was negatively regulated by the AAV *rep* gene products in COS-7 cells which constitutively express the SV40 T-antigen. Expression from the B19p6 promoter was also not detected in KB cells in the absence of Ad2, presumably because the AAV Rep proteins down-regulate expression from this promoter. However, expression from the B19p6 promoter was unaffected by the AAV Rep proteins in primary human hematopoietic progenitor cells in our studies. These studies suggest, therefore, that the erythroid cell-specific transcription factors, undoubtedly recruited by B19 to facilitate expression of the viral genes, predominate over the negative influence of the AAV Rep proteins. The precise role of these erythroid transcription factors in the B19 lifecycle is currently under investigation.

5 Summary and Conclusions

From the foregoing discussion, it is clear that a number of key elements in the overall biology of AAV still need to be studied before the full potential of this unique vector system can be realized. Although it is evident that extensive studies are currently underway in a number of laboratories, efforts in our own laboratory have focused on the molecular details of the AAV assembly, viral interaction with primary human hematopoietic cells, and the erythroid lineage-restricted expression of the transduced genes. We have identified the putative AAV packaging sequence, with which a host cell protein interacts specifically, and obtained preliminary evidence that AAV infection is mediated by a host cell receptor. We have also documented that it is indeed feasible to obtain erythroid lineage-specific expression of transduced genes using the AAV-B19 hybrid viral vector.

6 Future Prospects

An understanding of the molecular mechanisms underlying preferential replication and packaging of the AAV genomes should facilitate the production of high titer, defective-interfering particle-free, recombinant AAV containing therapeutic genes. The development of efficient AAV packaging cell lines remains a high priority as well. The availability of a homogeneous permanent cell line that lacks the putative receptor for AAV, and the possibility that this putative receptor could be introduced into these cells, should facilitate further studies on virus-host cell interactions. The identification and characterization of the putative receptor for AAV also has important implications in the potential use of AAV-based vectors in human gene therapy. Finally, the development of the AAV-B19 hybrid vector system promises to lead to potential gene therapy of human hemoglobinopathies in general and sickle cell anemia and β-thalassemia in particular (Zhou et al. 1995).

Acknowledgements. We thank Drs. Kenneth I Berns, Hal E. Broxmeyer and David A. Williams for their encouragement and support. We also thank Dr. Doris A. Morgan for generously providing the MB-02 cell line, and Drs. Peter J. Tattersall and Richard J. Samulski for their kind gifts of the recombinant B19 and AAV plasmids, respectively.

The research in the authors' laboratory was supported in part by grants from the National Institutes of Health (Public Health Service grants AI-26323, HL-48342, DK-49218 and HL-53586), and the Phi Beta Psi Sorority. A.S. is an Established Investigator of the American Heart Association.

References

Ashktorab H, Srivastava A (1989) Identification of nuclear proteins that specifically interact with adeno-associated virus type 2 inverted terminal repeat hairpin DNA. J Virol 63: 3034–3039

Berns KI (1990) Parvovirus replication. Microbiol Rev 54: 316–329

Berns KI, Bohenzky RA (1987) Adeno-associated viruses: an update. Adv Virus Res 32: 243–306

Brown KE, Anderson SM, Young NS (1993) Erythrocyte P antigen: cellular receptor for B19 parvovirus. Science 262: 114–117

Brown KE, Young NS, Liu JM (1994) Molecular, cellular and clinical aspects of parvovirus B19 infection. Crit Rev Oncol Hematol 16: 1–31

Carter BJ (1993) Adeno-associated virus vectors. Curr Opin Biotechnol 3: 533–538

Cotmore SF, Tattersall PJ (1984) Characterization and molecular cloning of a human parvovirus genome. Science 226: 1161–1165

Curiel DT, Wagner E, Cotten M, Birnstiel ML, Agarwal S, Li C-M, Hu P-C (1992) High-efficiency gene transfer mediated by adenovirus coupled to DNA-polylysine complexes. Hum Gene Ther 3: 147–154

Gottlieb J, Muzyczka N (1988) In vitro excision of adeno-associated virus DNA from recombinant plasmids: isolation of an enzyme fraction from HeLa cells that cleaves DNA at poly(G) sequences. Mol Cell Biol 8: 2513–2522

Hermonat PL, Labow MA, Wright R, Berns KI, Muzyczka N (1984) Genetics of adeno associated virus: isolation and preliminary characterization of adeno-associated virus type 2 mutants. J Virol 51: 329–339

Hirt B (1967) Selective extraction of polyoma DNA from infected mouse cultures. J Mol Biol 26: 365–369

Holliday R (1964) A mechanism for gene conversion in fungi. Genet Res 5: 282–304

Hong G, Ward P, Berns KI (1992) In vitro replication of adeno-associated virus DNA. Proc Natl Acad Sci USA 89: 4673–4677

Im D-S, Muzyczka N (1989) Factors that bind to adeno-associated virus terminal repeats. J Virol 63: 3095–3104

Im D-S, Muzyczka N (1990) The AAV origin binding protein Rep68 is an ATP-dependent site-specific endonuclease with DNA helicase activity. Cell 61: 447–457

Im D-S, Muzyczka N (1992) Partial purification of adeno-associated virus Rep78, Rep52, and Rep40 proteins and their biochemical characterization J Virol 66: 1119–1128

Kotin RM, Berns KI (1989) Organization of adeno-associated virus DNA in latently infected Detroit 6 cells. Virol 170: 460–467

Kotin RM, Siniscalco M, Samulski RJ, Zhu X, Hunter L, Laughlin CA, McLaughlin S, Muzyczka N, Rocchi M, Berns KI (1990) Site-specific integration by adeno-associated virus. Proc Natl Acad Sci USA 87: 2211–2215

Kotin RM, Menninger JC, Ward DC, Berns KI (1991) Mapping and direct visualization of a region-specific viral DNA integration site on chromosome 19q13-qter. Genomics 10: 831–834

Kotin RM, Linden RM, Berns KI (1992) Characterization of a preferred site on human chromosome 19q for integration of adeno-associated virus DNA by non-homologous recombination. EMBO J 11: 5071–5078

Labow MA, Berns KI (1988) The adeno-associated virus rep gene inhibits replication of an adeno-associated virus/simian virus 40 hybrid genome in cos-7 cells. J Virol 62: 1705–1712

Liu JM, Fujii H, Green SW, Komatsu N, Young NS, Shimada T (1991) Indiscriminate activity from the B19 parvovirus p6 promoter in non-permissive cells. Virol 182: 361–364

Liu JM, Green SW, Shimada T, Young NS (1992) A block in full-length transcript maturation in cells non-permissive for B19 parvovirus. J Virol 66: 4686–4692

Lusby E, Fife KH, Berns KI (1980) Nucleotide sequence of the inverted terminal repetition in adeno-associated virus DNA. J Virol 34: 402–409

Marcus DM, Kundu SK, Suzuki A (1981) The P blood group system: recent progress in immunochemistry and genetics. Semin Hematol 18: 63–71

Moritz T, Patel VK, Williams DA (1994) Bone marrow extracellular matrix molecules improve gene transfer into human hematopoietic cells via retroviral vectors. J Clin Invest 93: 1451–1457

Munshi NC, Zhou SZ, Woody MJ, Morgan DA, Srivastava A (1993) Successful replication of parvovirus B19 in the human megakaryocytic leukemia cell line MB-02. J Virol 67: 562–566

Muzyczka N (1992) Use of adeno-associated virus as a general transduction vector for mammalian cells. In: Muzyczka N (ed) Viral expression vectors. Springer, Berlin Heidelberg New York, pp 97–129 (Current topics in microbiology and immunology, vol 158)
Nahreini P, Srivastava A (1989) Rescue and replication of the adeno-associated virus 2 genome in mortal and immortal human cells. Intervirology 30: 74–85
Nahreini P, Srivastava A (1992) Rescue of the adeno-associated virus 2 genome correlates with alterations in DNA-modifying enzymes in human cells. Intervirology 33: 109–115
Nahreini P, Woody MJ, Zhou SZ, Srivastava A (1993) Versatile adeno-associated virus 2-based vectors for constructing recombinant virions. Gene 124: 257–262
Ni T-H, Zhou X-H, McCarty DM, Zolotukhin I, Muzyczka N (1994) In vitro replication of adeno-associated virus DNA. J Virol 68: 1128–1138
Owens RA, Carter BJ (1992) In vitro resolution of adeno-associated virus DNA hairpin termini by wild-type Rep protein is inhibited by a dominant-negative mutant of Rep. J Virol 66: 1236–1240
Ozawa K, Kurtzman GJ, Young NS (1986) Replication of the B19 parvovirus in human bone marrow cultures. Science 233: 883–886
Ozawa K, Kurtzman GJ, Young NS (1987) Productive infection by B19 parvovirus of human erythroid bone marrow cells in vitro. Blood 70: 384–391
Ponnazhagan S, Nallari ML, Srivastava A (1994) Suppression of human α-globin gene expression mediated by the recombinant adeno-associated virus 2-based antisense vectors. J Exp Med 179: 733–738
Ponnazhagan S, Wang XS, Woody MJ, Luo F, Kang LY, Nallari ML, Munshi NC, Zhou SZ, Srivastava A (1996) Differential expression in human cells from the p6 promoter of human parvovirus B19 following plasmid transfection and recombinant adeno-associated virus 2 (AAV) infection: human megakaryocytic leukemia cells are non-permissive for AAV infection. J Gen Virol (in press)
Rouger P, Gane P, Salmon C (1987) Tissue distribution of H, Lewis and P antigens shown by a panel of 18 monoclonal antibodies. Rev Fr Trans Immunohematol 30: 699–708
Sambrook J, Fritsch EF, Maniatis T (1989) Molecular cloning: a laboratory manual. Cold Spring Harbor Laboratory Press, Cold Spring Harbor, pp 1.53–1.110
Samulski RJ, Berns KI, Tan M, Muzyczka N (1982) Cloning of adeno-associated virus into pBR322: rescue of intact virus from the recombinant plasmid in human cells. Proc Natl Acad Sci USA 79: 2077–2081
Samulski RJ, Srivastava A, Berns KI, Muzyczka N (1983) Rescue of adeno-associated virus from recombinant plasmids: gene correction within the terminal repeats of AAV. Cell 33: 135–143
Samulski RJ, Chang L-S, Shenk T (1987) A recombinant plasmid from which an infectious adeno-associated virus genome can be excised in vitro and its use to study viral replication. J Virol 61: 3096–3101
Samulski RJ, Chang L-S, Shenk T (1989) Helper-free stocks of recombinant adeno-associated viruses: Normal integration does not require viral gene expression. J Virol 63: 3822–3828
Samulski RJ, Zhu X, Xiao X, Brook J, Houseman DE, Epstein N, Hunter LA (1991) Targeted integration of adeno-associated virus (AAV) into human chromosome 19. EMBO J 10: 3941–3950
Schwarz TF, Serke S, Hottentrager B, von Brunn A, Baurmann H, Kirsch A, Stolz W, Hunn D, Deinhardt F, Roggendrof M (1992) Replication of parvovirus B19 in hematopoietic progenitor cells generated in vitro from normal human peripheral blood. J Virol 67: 562–566
Senapathy P, Tratschin J-D, Carter BJ (1984) Replication of adeno-associated virus DNA. Complementation of naturally occurring rep mutants by a wild-type genome or an ori mutant and correction of terminal palindrome deletions. J Mol Biol 179: 1–20
Shade RO, Blundell MC, Cotmore SF, Tattersall PJ, Astell CR (1986) Nucleotide sequence and genome organization of human parvovirus B19 isolated from the serum of a child during aplastic crisis. J Virol 58: 921–936
Shimomura S, Komatsu N, Frickhofen N, Anderson S, Kajigaya S, Young NS (1992) First continuous propagation of B19 parvovirus in a cell line. Blood 79: 18–24
Siegl G, Bates RC, Berns KI, Carter BJ, Kelly DC, Kurstak E, Tattersall P (1985) Characteristics and taxonomy of parvoviridae. Intervirology 23: 61–73
Snyder RO, Samulski RJ, Muzyczka N (1990) In vitro resolution of covalently joined AAV chromosome ends. Cell 60: 105–113
Snyder RO, Im D-S, Ni T-H, Xiao X, Samulski RJ, Muzyczka N (1993) Features of the adeno-associated virus origin involved in substrate recognition by the viral Rep protein. J Virol 67: 6096–6104
Spalholz BA, Tattersall PJ (1983) Interaction of minute virus of mice with differentiated cells: strain dependent target-cell specificity is mediated by the intracellular factors. J Virol 46: 937–943

Srivastava A (1987) Replication of the adeno-associated virus DNA termini in vitro. Intervirology 27: 138–147

Srivastava A (1994) Parvovirus-based vectors for human gene therapy. Blood Cells 20: 531–538

Srivastava A, Lu L (1988) Replication of B19 parvovirus in highly enriched hematopoietic progenitor cells from normal human bone marrow. J Virol 62: 3059–3063

Srivastava A, Lusby EW, Berns KI (1983) Nucleotide sequence and organization of the adeno-associated virus 2 genome. J Virol 45: 555–564

Srivastava A, Bruno E, Briddell R, Cooper R, Srivastava CH, van Besien K, Hoffman R (1990) Parvovirus B19-induced perturbation of human megakaryocytopoiesis in vitro. Blood 76: 1997–2004

Srivastava CH, Samulski RJ, Lu L, Larsen SH, Srivastava A (1989) Construction of a recombinant human parvovirus B19: Adeno-associated virus 2 (AAV) DNA inverted terminal repeats are functional in an AAV-B19 hybrid virus. Proc Natl Acad Sci USA 86: 8078–8082

Srivastava CH, Zhou SZ, Munshi NC, Srivastava A (1992) Parvovirus B19 replication in human umbilical cord blood cells. Virology 189: 456–461

Takahashi T, Ozawa K, Takahashi K, Asano S, Takaku F (1990) Susceptibility of human erythropoietic cells to B19 parvovirus in vitro increases with differentiation. Blood 75: 603–610

Tratschin J-D, Miller IL, Carter BJ (1984) Genetic analysis of adeno-associated virus: properties of deletion mutants constructed in vitro and evidence for an adeno-associated virus replication function. J Virol 51: 611–619

Tuan DYH, Solomon WB, London IM, Lee DP (1989) An erythroid specific development stage-independent enhancer far upstream of the human "β-like globin " genes. Proc Natl Acad Sci USA 86: 2554–2559

von dem Borne AEGK, Bos MJE, Joustra-Maas N, Tromp JF, van Wijngaarden-du Bois R, Tetteroo PAT (1986) A murine monoclonal IgM antibody specific blood group P antigen (globoside). Br J Haematol 63: 35–46

Walsh CE, Liu JM, Xiao X, Young NS, Nienhuis AW, Samulski RJ (1992) Regulated high level expression of a human γ globin gene introduced into erythroid cells by an adeno-associated virus vector. Proc Natl Acad Sci USA 89: 7257–7261

Wang X-S, Ponnazhagan S, Srivastava A (1995a) Rescue and replication signals of the adeno-associated virus 2. J Mol Biol 250: 573–580

Wang X-S, Ponnazhagan S, Srivastava A (1995b) Rescue and replication of adeno-associated virus type 2 genome as well as vector DNA sequences from recombinant plasmids containing deletions in viral inverted terminal repeats: selective encapsidation of viral genomes in progeny virions. J Virol 70: 1668–1677

Wang X-S, Yoder MC, Zhou SZ, Srivastava A (1995c) Parvovirus B19 promoter at map unit 6 confers replication-competence and erythroid-specificity to adeno-associated virus 2 in primary human hematopoietic progenitor cells. Proc Natl Acad Sci USA 92: 12416–12420

Ward P, Berns KI (1991) In vitro rescue of an integrated hybrid adeno-associated virus/simian virus 40 genome. J Mol Biol 218: 791–804

Yaegashi N, Shiraishi Y, Takeshita T, Nakamura M, Yajima A, Sugamura K (1989) Propagation of human parvovirus B19 in primary culture of erythroid lineage cells derived from fetal liver. J Virol 63: 2422–2426

Zhou SZ, Broxmeyer HE, Cooper S, Harrington MA, Srivastava A (1993) Adeno-associated virus 2-mediated gene transfer in murine hematopoietic progenitor cells. Exp Hematol 21: 928–933

Zhou SZ, Cooper S, Kang LY, Ruggieri L, Heimfeld S, Srivastava A, Broxmeyer HE (1994) Adeno-associated virus 2-mediated high efficiency gene transfer into immature and mature subsets of hematopoietic progenitor cells in human umbilical cord blood. J Exp Med 179: 1867–1875

Zhou SZ, Li Q, Stamatoyannopoulos G, Srivastava A (1996) Adeno-associated virus 2-mediated transduction and erythroid cell-specific expression of a human β-globin gene. Gene Therapy 3: 223–229

Development of Adeno-associated Virus Vectors for Gene Therapy of Cystic Fibrosis

B.J. Carter[1] and T.R. Flotte[2]

[1]Targeted Genetics Corporation, 1100 Olive Way, Suite 100, Seattle, WA 98101, USA
[2]Eudowood Division of Pediatric Respiratory Sciences, Johns Hopkins University School of Medicine, 600 North Wolfe Street, Baltimore, MD 21287–2533, USA

1 Adeno-associated Virus Vectors

The potential application of adeno-associated virus (AAV) as a gene delivery system in mammalian cells was originally indicated by several features of its life cycle. Most importantly, in the absence of a helper virus for productive AAV replication, the AAV genome can persist in host cells and this may involve integration into the host cell genome. Appreciation of the technical feasibility of this potential gene therapy application arose from cloning infectious AAV genomes (SAMULSKI et al. 1982; LAUGHLIN et al. 1983). The earliest AAV vectors were reported about 10 years ago (TRATSCHIN et al. 1984; HERMONAT and MUZYCZKA 1984) but progress from there was slow because very few laboratories were engaged in this work. However, recent developments have highlighted the potential use of AAV vectors for gene therapy (FLOTTE et al. 1993a,b; KAPLITT et al. 1994) and AAV vectors are now being developed for imminent clinical trials in patients with cystic fibrosis (FLOTTE et al. 1993b, 1994a).

AAV vectors are constructed in AAV plasmids by substitution of foreign DNA for either or both of the *cap* and *rep* genes, leaving the inverted terminal repeats (ITRs) intact. These vectors can be packaged into AAV particles by complementation of the vector plasmid with a packaging system that expresses the *rep* and *cap* genes in adenovirus-infected cells. Vectors can then be purified and concentrated from cell-free lysates (CARTER 1992).

The initial AAV packaging systems were inefficient and generation of wild-type AAV by recombination was a problem (TRATSCHIN et al. 1985; HERMONAT and MUZYCZKA 1984; MCLAUGHLIN et al. 1988; LEBKOWSKI et al. 1988). Modifications to the system allowed generation of modest titers of AAV vectors that were relatively free of wild-type AAV (SAMULSKI et al. 1989). More recently, additional improvements have enhanced titers to usable levels and eliminated production of wild-type AAV (FLOTTE et al. 1995). Currently, a great deal of effort is being expended in developing maximally efficient, scalable packaging systems for AAV vectors.

AAV vectors are attractive from the point of view of safety because AAV has not been associated with any disease (BLACKLOW 1985). Also, it was reported that wild-type AAV preferentially integrates into a specific region of human chromosome 19 (KOTIN and BERNS 1989; KOTIN et al. 1990; SAMULSKI et al. 1991). However, this site preference for integration has been lost in AAV vectors that are deleted for AAV coding sequences (SAMULSKI 1993; W. Kearns et al. 1994, unpublished observations).

Latent AAV or AAV vectors that persist in host cells either as an episome or as an integrated proviral genome may be rescued and mobilized by superinfection with a helper adenovirus or perhaps a herpesvirus, but rescue also requires the AAV *rep* gene and packaging requires the *cap* gene (CARTER 1992; MUZYCZKA 1992). AAV vectors that do not contain these genes are not likely to be mobilized unless the cells are also coinfected with wild-type AAV in addition to the helper adenovirus. This emphasized the importance of developing AAV packaging systems that do not generate wild-type AAV (FLOTTE et al. 1995) and also indicated the need to assess the likelihood of adventitious AAV in human populations before AAV vectors were used in gene therapy protocols. One study using PCR analysis reported the detection of AAV2 sequences in peripheral blood from two of 55 healthy donors and two of 16 hemophiliac patients (GROSSMAN et al. 1992). Other studies have failed to detect any AAV sequences in 36 samples of lung tissue from CF patients (R.J. Samulski 1994, personal communication) or in paired samples of normal and tumor tissue from lung cancer patients (T.R. Flotte and S.A. Afione 1993, unpublished experiments).

In addition to the generic AAV vector issues of titer and freedom from wild-type AAV, several considerations were important in developing AAV vectors for a potential gene therapy of cystic fibrosis (CF). First, efficient transduction is required in vivo in the airway. Second, the CF transmembrane regulator (CFTR) cDNA coding sequence is almost too large to fit in an AAV vector that can be packaged into AAV particles. This was a critical issue in vector design and led to a consideration of how to obtain a very compact transcriptional regulatory system that was nevertheless functional in airway epithelial cells.

We developed AAV vectors for CF in several steps. (1) We first examined the function of AAV vectors in airway epithelial cells. (2) We then analyzed the ability of AAV vectors expressing CFTR to correct the physiologic defect in CF cells in vitro. (3) We also addressed the issue of specific vector design because of the size constraints of the CTFR cDNA in relation to the packaging constraints of AAV vectors. (4) We then addressed an in vivo model of delivery of AAV vectors to the lungs of rabbits to assess vector transduction efficiency, expression and persistence of the vector and to gain an initial assessment of safety and toxicity. (5) We then performed preclinical toxicity studies in rhesus macaques to determine the efficiency of delivery and the duration of expression and to evaluate safety and toxicity in the context of parameters that would be measured in human clinical trials. In the rhesus model we also analyzed the possibility of vector mobilization or shedding. (6) Finally, we modified the procedure for production of AAV to ensure production of sufficient vector for clinical trials and to avoid any contamination with wild-type AAV. The vector for the initial clinical trials will be made by transfecting adenovirus-infected 293 cells with the vector plasmid and a *rep-cap* complementing packaging plasmid. We have used an alternate complementing plasmid, pRS5, which expresses the *rep* gene from a heterologous promoter and which leads to enhanced yields of vector particles (FLOTTE et al. 1995).

2 Gene Therapy for Cystic Fibrosis

The airway epithelium is the critical site of cellular dysfunction in CF, the most common lethal genetic disease in North America (BOAT et al. 1989). The disease results from a mutation in the gene for the CFTR (RIORDAN et al. 1989). Therefore it may be possible to treat this disease by delivering a functional CFTR gene directly to the epithelial cells of the airway surface. The CF defect can be complemented by introduction and expression of the CFTR gene in human CF cells in culture (DRUMM et al. 1990; RICH et al. 1990). This potentially provides a valuable tool for both experimental and therapeutic applications. However, any in vivo gene delivery system for treatment of CF patients requires the capacity to transfer the gene to a high proportion of the cells and preferably to maintain appropriate levels of expression over prolonged periods.

One approach to in vivo gene therapy is to use viruses since they are currently the most efficient gene delivery systems. There may also be an advantage to using viruses that are tropic to airway epithelial cells (ROSENFELD et al. 1991). Unfortunately, retroviral vectors are of little utility because they do not function in nondividing cells (MILLER et al. 1990), which constitute much of the respiratory epithelium. Of the many other viruses that may infect the human respiratory epithelium, most have little or no potential to exhibit stable persistence or genome integration, and almost all are pathogenic to the host. One exception is AAV which, when used as a vector, can yield efficient transduction of cells. Second, AAV is known to be able to persist in cells for prolonged periods. Also, AAV has never been associated with causing any disease in humans or other animal species. Other viruses, such as adenoviruses, can efficiently deliver genes to airway cells in vivo (ROSENFELD et al. 1991). However, these viruses are naturally pathogenic in humans. In initial preclinical and clinical tests for CF gene therapy, adenovirus vectors have exhibited significant clinical toxicities in both nasal epithelium and lung (ALTON and GEDDES 1995; CRYSTAL et al. 1994; CRYSTAL and MCELVANEY 1995; ZABNER et al. 1993).

3 Expression from Adeno-associated Virus Vectors in Airway Epithelial Cells

We performed preliminary studies to determine if AAV2 vectors were likely to be useful for stable, high-level expression of genes introduced into a CF airway epithelial cell line by either transfection or virus particle-mediated transduction (FLOTTE et al. 1992). The size of the CFTR coding region (4.4 kb) is close to the packaging limit for AAV transducting vectors and, therefore, construction of an AAV-CFTR vector required very compact transcription regulatory elements. Initially, we analyzed the suitability of the AAV p5 promoter for expression because this promoter forms a convenient 100 bp cassette with the AAV left-hand ITR and is one of the smallest promoters available. We therefore analyzed reporter gene expression from the AAV p5 promoter in the IB3 cell line (FLOTTE et al. 1992). IB3 cells are immortalized human bronchial epithelial cell line derived from a CF patient (LUO et al. 1989; ZEITLIN et al. 1991). These cells retain characteristics of airway epithelial cells and are deficient in protein kinase A-mediated activation of chloride conductance, which is diagnostic of the CF genetic defect.

3.1 Activity of the AAV p5 Promoter for *cat* Gene Expression

We constructed the AAV-p5cat plasmid pRO1472 (Flotte et al. 1992) which contains the *cat* gene coding sequence positioned immediately downstream of AAV nucleotides 1 to 321. This region of AAV includes several notable features including the AAV replication origin (ITR) and the p5 promoter TATA box at nucleotide 255. Between these two regions there are several other transcription factor binding sites including an MLTF (USF) site, a YY1 site and a site between nucleotides 160 and 180, comprising the sequence GTGACGTGAATTACGTCA-TAG, which has homology to the cAMP response element (CRE) and the binding site for the CREB/ATF transcription factor family (Hai et al. 1988; Lin and Green 1988; Montminy and Bilezkjian 1987; Montminy et al. 1990; Sassoni-Corsi 1988). Several additional AAVp5 cat plasmids were derived that had the ITR or CRE deleted.

The AAVp5cat plasmids were tested for CAT expression in transient assays after transfection into IB3 cells (Flotte et al. 1992). The plasmid pR01472, containing both the ITR and the entire AAV p5 promoter, had the highest activity in IB3 cells. Furthermore, the activity of the complete AAV p5 promoter was more than ten fold higher than that of the SV40 early promoter in IB3 cells. To determine if the CRE might function in the context of the complete AAV p5 promoter, we tested the effect of forskolin-induced elevation of the level of cAMP on the subsequent expression of CAT. In IB3 cells, there was a two fold increase of CAT activity from the complete ITR-p5 promoter in response to forskolin but there was no effect if the CRE was deleted.

3.2 Activity of the AAV p5 Promoter for *neo* Gene Expression

To test the p5 promoter for longer term expression, we constructed a pAAVp5neo vector (Flotte et al. 1992). This plasmid contains a *neo* gene coding sequence inserted immediately downstream from the p5 promoter and left-hand ITR, analogous to the AAVp5 CAT plasmid pRO1472. The plasmid pSVneo contains *neo* expressed from an SV40 early gene promoter and the plasmid pAAVSVneo contains the SV40 promoter and *neo* gene cassette flanked by left and right AAV ITRs. The plasmids pAAVp5neo, pSVneo, and pAAVSVneo were compared by transfecting them into IB3-1 cells using lipofectin followed by selection of cells in the antibiotic geneticin.

Both pAAVp5neo and pAAVSVneo transfections showed seven to eight fold increase over the number of colonies produced by pSV2neo. This suggested that in IB3 cells the AAV ITR can function as an enhancer for the SV40 promoter. A similar observation regarding a possible enhancer role for the ITRs in human HeLa cells was reported by Beaton et al. (1989). These experiments showed that the AAV p5 promoter was relatively efficient for expression of a selective marker in the IB3 cell line.

3.3 Gene Transfer by AAV *neo* Transducing Particles

The above experiments showed that the AAV p5 promoter in the presence of the AAV ITR functioned for transient gene expression as well as for stable expression of the *neo* gene in IB3 cells. However, delivery of plasmid DNA by lipofection is generally not an efficient gene delivery system in mammalian cells. In contrast, AAV vector DNA can be packaged into AAV particles, and infection of cells with these AAV transducing particles is a much more efficient delivery system in several human cell lines (HERMONAT and MUZYCZKA 1994; TRATSCHIN et al. 1985). To test this in human CF airway epithelial cells, we first packaged the AAVp5neo vector into AAV transducing particles by cotransfecting adenovirus-infected 293 cells with the vector plasmid, pAAVp5neo DNA, together with the packaging plasmid pAAV/Ad DNA (SAMULSKI et al. 1989) to provide complementing *rep* and *cap* functions (FLOTTE et al. 1992). The AAVp5neo transducing particles were concentrated from cell lysates by CsCl equilibrium density gradient centrifugation and used for transduction of IB3 cells.

Transduction of up to 60%–70% of the IB3 cells with an input of about 850 transducing particles per cell, which we achieved with the AAV vector in IB3 cells (FLOTTE et al. 1992), compared well with an earlier report of transduction of 70% of the cells by an AAVSV-neo transducing vector in human Detroit 6 cells (SAMULSKI et al. 1989). In both studies, a multiplicity of up to 800–1000 particles was required to reach this level. These studies showed that AAV vectors can yield high transduction efficiencies and that , in a CF airway epithelial cell line, the AAV p5 promoter functioned efficiently in the transducing vector.

3.4 Expression of Reporter Genes Using the Inverted Terminal Repeat as a Promoter

The p5 promoter is small and is contained as a convenient 263–321 bp cassette together with the AAV ITR, which must in any case be included in any AAV transducing vector. However, in subsequent work (FLOTTE et al. 1993a) we observed serendipitously that the AAV ITR, which is only 145 bp in length, was apparently able to influence gene expression by functioning directly as a transcription promoter. We confirmed this novel observation by constructing an AAV vector in which the reporter gene, *cat*, was linked directly to the ITR and was expressed when introduced into cells. This observation proved useful in constructing AAV-transducing vectors for expression of CFTR containing a minimally sized promoter.

We compared the AAV-CAT vector plasmid pRO1472, which has the complete ITR and AAV p5 promoter upstream of the CAT gene, with a plasmid, pTRF46, in which the CAT gene was inserted immediately following the AAV ITR sequence. These AAV-CAT plasmids were transfected into human (293) cells, and CAT activity was measured in extracts prepared 48 h later. Surprisingly, pTRF46 expressed CAT at levels almost one half of pRO1472, even though it contains none of the previously characterized p5 promoter elements (i.e., nucleotides 145–320). This was a novel observation which showed that the AAV ITR sequence itself is capable of acting as an efficient promoter for gene expression (FLOTTE et al. 1993a).

4 In Vitro Studies with Adeno-associated Virus Cystic Fibrosis Transmembrane Regulator Vectors

4.1 Expression of CFTR with AAV Vectors Complements the Cystic Fibrosis Defect in Stable Transfectants of Cystic Fibrosis Airway Cells

To examine expression of the CFTR cDNA, we constructed several AAV-CFTR vector plasmids (Fig. 1). CFTR was expressed either from the AAV p5 promoter, in pSA313, or from the AAV ITR, in pSA315, pSA306 and pTRF42. In pSA313, the CFTR cDNA of 4502 nucleotides was inserted downstream of an AAV promoter (i.e., nucleotides 1–263). In pSA315 the CFTR cDNA was inserted in the opposite orientation such that it is downstream of the right-hand AAV ITR sequence and an inverted copy of a 60 bp synthetic poly (A) site. In this configuration the CFTR cDNA is expressed from the right-hand ITR. The pSA306 construct was exactly analogous to pSA315, except that 363 nucleotides of the NH_2 terminal region of CFTR cDNA were deleted. This resulted in expression from the right-hand ITR of a fusion protein consisting of an NH_2-terminally deleted CFTR protein having a novel region at its NH_2 terminal, derived from reading through the synthetic poly (A) site in the reverse direction, fused in-frame with the CFTR cDNA beginning with amino acid I119 in the second transmembrane domain. The plasmid pSA464 was derived from pSA306 by introducing a frameshift mutation such that it cannot produce a functional CFTR protein. pTRF42 contained only the AAV ITR promoter sequence to nucleotide 145 and the CFTR sequence beginning at nucleotide 486, without any additional fusion, except for an in-frame ATG for initiation of translation.

The AAV-CFTR plasmids were each transfected into IB3 cells, together with pAAVp5neo (FLOTTE et al. 1993a). Control cells were transfected with pAAVp5-neo alone. Geniticin-resistant colonies were expanded into stable cultures and characterized for functional expression of the CFTR protein.

Expression of CFTR can be detected in functional assays in IB3 cells which have the CF defect. A functional CFTR protein should restore to these cells a Cl^- conductance which is regulated by cAMP and thus is stimulated by forskolin (HWANG et al. 1989; LI et al. 1988; DRUMM et al. 1990; RICH et al. 1990). This can be measured in a Cl^- efflux assay by measuring the rate of efflux of radioactive Cl^- when cells are shifted to Cl^- free medium. Both the parental IB3 cells and a

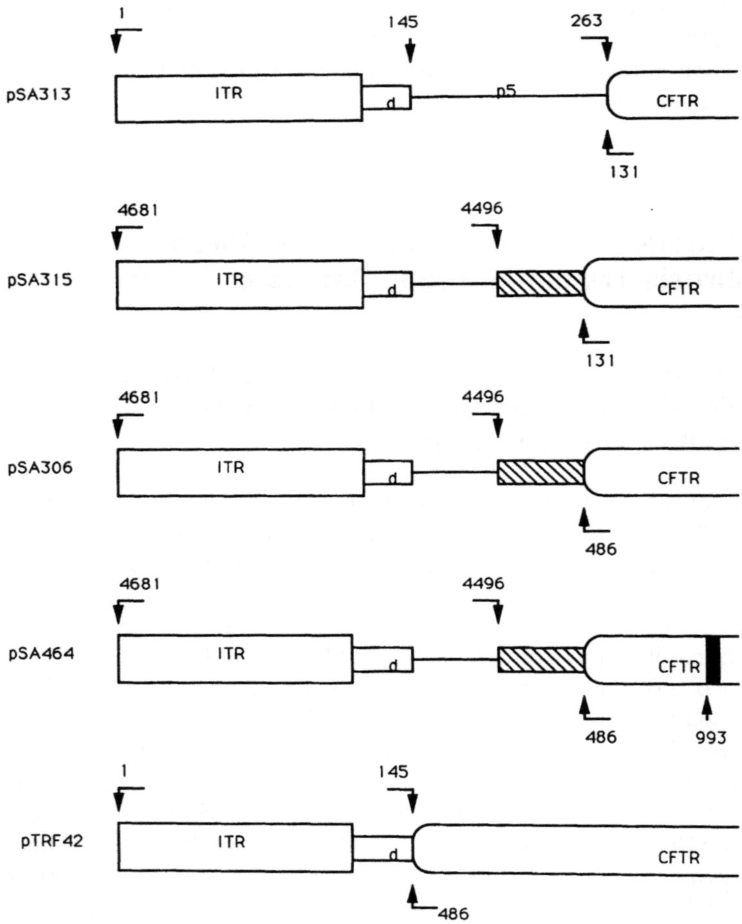

Fig. 1. The structure of the promoter regions of adeno-associated virus-cystic fibrosis transmembrane regulator vectors. pSA313 contains the CFTR cDNA inserted downstream of the AAV left-hand inverted terminal repeat (ITR) and the p5 promoter (i.e., at AAV nucleotide 263) and contains a synthetic poly A site downstream of the CFTR insert (not shown). pSA315 has the same CFTR insert as pSA313 in the same plasmid, but the inserted CFTR cDNA is inserted in the opposite direction and is expressed from the right-hand AAV ITR promoter. pSA306 is the same as pSA313 except for a deletion of nucleotides 123–486 from the CFTR sequence. pSA464 is the same as pSA306 except that a frameshift mutation (*vertical solid bar*) was introduced at an *Afl* II site at nucleotide 993 in the CFTR sequence. pTRF42 contains a CFTR insert analogous to that in pSA306 but which is expressed directly from the AAV ITR. AAV2 nucleotide numbers are indicated above each diagram, whereas CFTR cDNA numbering is indicated below. The *hatched bar* indicates the 60-base synthetic nucleotide which contains a poly A sequence. Note that in pSA315, pSA306 and pSA464, this sequence is inserted in the reverse orientation. For simplicity, the pBR322 sequences and the poly A or ITR sequences downstream of the CFTR insert are not shown. (From FLOTTE et al. 1993a)

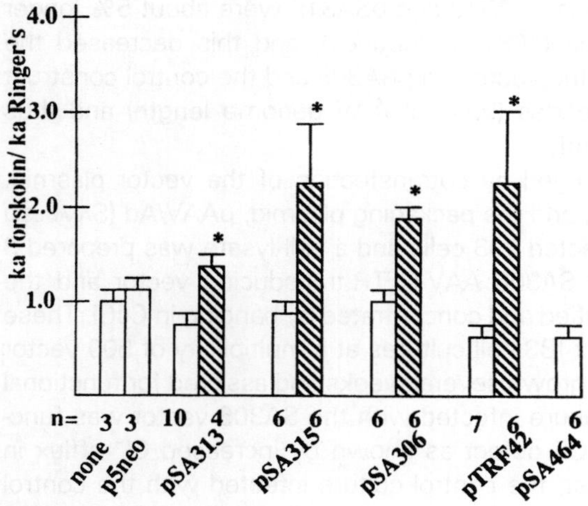

Fig. 2. Chloride efflux assays in IB3-1 cells complemented with adeno-associated virus-cystic fibrosis transmembrane regulator vectors. IB-3 cells were transfected with pAAVp5neo and either pSA464, pSA313, pSA315, pSA306, or pTRF42. Geneticin-resistant clones were selected and analyzed for responsiveness to forskolin stimulation in a Cl⁻ efflux assay. The ratio of the rate of efflux in the presence of forskolin to the rate in the absence of forskolin (k_a forskolin/k_a Ringer's) is plotted. For each vector, *n* indicates the number of individual clones which did (*hatched bars*) or did not (*open bars*) show a forskolin response. For each group of clones, the average ratio was calculated. (From Flotte et al. 1993a)

control clone N6 (transfected with pAAVp5neo alone) exhibited a relatively slow Cl^- efflux rate that was not responsive to forskolin (Fig. 2). In contrast , clones of AAV-CFTR transfectants exhibited an increased basal rate of efflux which was further stimulated with forskolin as expected for normal CFTR function. All four vector constructs including the p5 vector (pSA313) and the ITR vector (pSA315, pSA306, pTRF42) were functional. None of the clones transfected with the control vector pSA464 were complemented.

These results showed two novel findings. First, the AAV ITR sequence functioned efficiently as a promoter for CFTR expression as shown by the function of both pSA315 and pSA306. Second, the truncated CFTR proteins expressed from pSA306 or pTRF42 were also functional for complementation of the CFTR defect.

4.2 Packaged AAV-CFTR Transducing Vectors Complement the Cystic Fibrosis Defect in the Absence of Selective Pressure

The AAV CFTR plasmids containing either full length or the truncated CFTR coding sequence were functional in the stable transfection assays. To test these constructs as transducing vectors it was necessary to package them into AAV particles. The vectors contained in pSA313 and pSA315 were about 5% longer than the wild-type AAV genome (4681 nucleotides), and this decreased the packaging efficiency. However, the vectors in pSA306 and the control construct pSA464 were only 4647 nucleotides (99% of AAV genome length) and thus could be packaged more efficiently.

SA306 or SA364 was packaged by cotransfection of the vector plasmid, pSA306 or pSA464 respectively, and the packaging plasmid, pAAV/Ad (Samulski et al. 1989), into adenovirus-infected 293 cells and a cell lysate was prepared 3 days later. Preparations of the SA306 AAV-CFTR-transducing vector and the control vector, SA464, were purified and concentrated by banding in CsCl. These vectors were then used to infect IB3 cell cultures at a multiplicity of 500 vector particles/cell. The cultures were grown several weeks and assayed for functional expression of the CFTR. The culture infected with the SA306 vector was functionally complemented for the CF defect as shown by increased Cl^- efflux in response to forskolin. In contrast, the control culture infected with the control SA464 vector (2F2 cells) was not complemented as shown by the lack of response to forskolin (Flotte et al. 1993a).

The chloride efflux studies were consistent with functional complementation of the CF defect, but the efflux technique does not definitively identify the actual channel responsible for the complementation. For this reason we also performed electrophysiologic studies on isolated, cell-detached patches from bulk cultures and individual clones. This confirmed the functional complementation (Egan et al. 1992) and showed that it was due to the CFTR channel. Also these studies resolved an additional feature of the chloride channel defect in CF.

4.3 Defective Regulation of Outwardly Rectifying Chloride Channels by Protein Kinase A Corrected by Insertion of CFTR

The CFTR protein functions as a low conductance Cl^- channel with a linear current-voltage relationship whose regulation is defective in CF patients (Kartner et al. 1991; Anderson et al. 1991; Bear et al. 1991). Larger conductance, outwardly rectifying Cl^- channels are also defective in CF and fail to activate when exposed either to cyclic AMP-dependent protein kinase A or to protein kinase C (Li et al. 1988; Schoumacher et al. 1987; Chen et al. 1989; Hwang et al. 1989). However, the role of the outwardly rectifying Cl^- channel in CF had been questioned (Ward et al. 1991). In addition no relationship between this channel and the CFTR chloride channel had been demonstrated. Expression of recombinant CFTR using AAV vectors in CF bronchial epithelial cells corrected defective Cl^- secretion, in that it induced the appearance of small, linear Cl^- conductance channels, but surprisingly, it also restored protein kinase A activation of the outwardly rectifying Cl^- channels (Egan et al. 1992). These results clarified the involvement of outwardly rectifying Cl^- channels in CF and suggest that CFTR regulates more than one conductance pathway in airway tissues.

In these studies by EGAN et al. (1992) excised patch-clamp experiments were performed on IB3 cells that had been transfected with AAV-CFTR plasmids or that had been infected with AAV-CFTR viral particles. The AAV-CFTR plasmids (pSA315 or pSA306) were cotransfected into IB3 cells with the AAV*neo* plasmid to produce geneticin-resistant clones and screened in Cl^- efflux assay for forskolin-dependent stimulation of Cl^- secretion. Transfection with the plasmid pSA464, having a frame-shift mutation, gave no complemented clones. When the AAV-CFTR vector pSA306 was packaged into AAV particles and infected into IB3 cells, cAMP regulation of Cl^- efflux was also complemented in the mass culture. Cells transduced with AAV particles containing the frame-shift mutant vector were not complemented.

No spontaneous channel activity occurred when patches were excised from IB3 cells and addition of protein kinase A (PKA) and ATP 2 min after excision failed to activate channels (Fig. 3A). PKA also failed to activate Cl^- channels in IB3 cells infected with AAV virus particles containing CFTR complementary DNA with a frameshift mutation that cannot produce a functional CFTR protein. Two

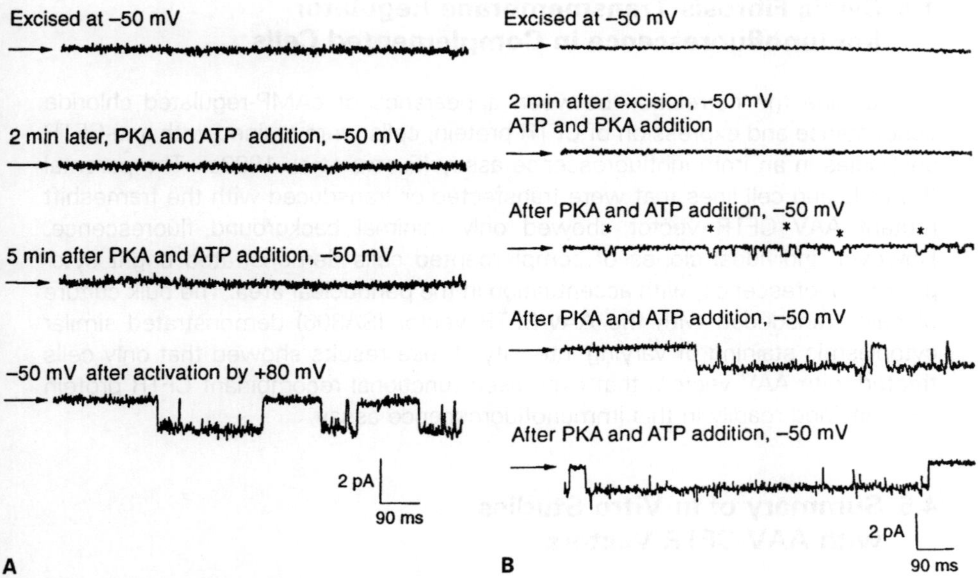

Fig. 3. A Chloride channels in cystic fibrosis bronchial epithelial cells (IB3-cells). Excised patches from IB3 cells were held at –50 mV for 2 min before addition of 50 n*M* protein kinase A (PKA) and 1 mM Mg-ATP to the cytoplasmic side of the membrane. The failure of PKA activation in this patch was monitored for an additional 5 min. Presence of outwardly rectifying chloride channels was verified by voltage activation (*bottom trace*). **B** PKA activation of low conductance and outwardly rectifying Cl^- channels in IB3 cells complemented with AAV-CFTR vectors. Patches were excised from complemented cells and held at –50 mV for at least 2 min before PKA and Mg-ATP were added (as in **A**). *Asterisks* indicate low conductance channel activity. In most experiments involving PKA activation, low conductance channels activated before outwardly rectifying Cl^- channels, although in some instance both channels could be observed to activate simultaneously. (From EGAN et al. 1992)

types of channels were identified in complemented IB3 cells (Fig. 3B). The addition of 50 n*M* PKA and 1 m*M* ATP activated both small, linear conductance channels characteristic of CFTR expression and also activated outwardly rectifying Cl^- channels identical to those observed in previous studies in normal airway cells (Hwang et al. 1989).

These experiments demonstrated that complementation with AAV-CFTR vectors corrected defective regulation of outwardly rectifying Cl^- channels and induced the expression of small, linear conductance channels charcteristic of the CFTR protein. How CFTR might control the PKA activation of outwardly rectifying Cl^- channels is not yet known but it is unlikely that the CFTR gene also encodes this channel. It is possible that CFTR may induce PKA regulation by affecting outwardly rectifying Cl^- channels as they are processed in the intracellular organelles. Alternatively, CFTR may associate with or transport a substance that is necessary for PKA activation of outwardly rectifying Cl^- channels (Barasch et al. 1991). Recent studies (Schwiebert et al. 1995) suggest that this substance might be ATP.

4.4 Cystic Fibrosis Transmembrane Regulator Immunofluorescence in Complemented Cells

To examine the correlation between appearance of cAMP-regulated chloride conductance and expression of CFTR protein, cells were stained with anti-CFTR antibodies in an immunofluorescence assay (Flotte et al. 1993a). The parental IB3 cells and cell lines that were transfected or transduced with the frameshift mutant AAV CFTR vector showed only minimal background fluorescence. However, individual clones of complemented cells demonstrated bright cytoplasmic fluorescence, with accentuation in the perinuclear area. The bulk culture of cells transduced with the AAV-CFTR vector (SA306) demonstrated similar cytoplasmic staining of varying intensity. These results showed that only cells treated with AAV vectors that expressed functional recombinant CFTR protein were stained readily in this immunofluorescence assay.

4.5 Summary of In Vitro Studies with AAV-CFTR Vectors

The in vitro studies described above revealed several novel observations. First, we showed that the AAV ITR sequence can function directly as a promoter for expression of a reporter gene in a transient assay or for expression of CFTR in a long term stable expression assay. Second, we showed that the CFTR cDNA expressed from either the AAV p5 promoter or the ITR could functionally complement the Cl^- conductance defect in airway epithelial cells from a CF patient. Third, we showed that an NH_2 terminally deleted CFTR protein could also complement the CF defect. This latter observation, together with the use of minimally sized AAV ITR promoter, facilitated packaging and functional expression from an AAV transducing vector, a gene whose coding region otherwise exceeded the packaging capacity of AAV. These observations have several important implications for understanding the life cycle of AAV (which will not be discussed here), for development of AAV vectors, and for understanding the function of the CFTR gene.

The observation that the AAV ITR could function directly as a promoter was unexpected since this region was not thought to behave as a transcriptional regulatory sequence (MUZYCZKA 1992; CARTER 1990; CARTER et al. 1990; WALSH et al. 1992). Also, deletion of the ITR from the AAV p5 promoter did not have a major effect on expression (FLOTTE et al. 1992). However, inspection of the ITR sequences showed several regions that correspond to Sp1 binding sites (MITCHELL and TJIAN 1989; BRIGGS et al. 1986; PITLUK and WARD 1991), and the AAV D-sequence (nucleotides 125–145) bears a strong homology to the consensus Inr sequence found at the transcriptional start site of many genes (SMALE and BALTIMORE (1989). Furthermore, the Inr sequence together Sp1 sites is capable of functioning as a promoter in the absence of a TATA motif (SMALE et al. 1990; DEPAMPHILIS 1988).

The AAV ITR cassette used as a promoter represents the minimal sequence potentially capable of allowing for viral DNA replication, packaging, and gene expression. This effectively expands the space available for foreign DNA inserts in AAV vectors by reducing the viral DNA sequence requirement in the vector to a total of 290 nucleotides. It also eliminated the need for adding additional promoter sequences and so saved up to several hundred nucleotides of sequence. This may be a crucial difference for successful construction of some AAV vectors such as the AAV-CFTR vectors.

The in vitro studies with the AAV-CFTR vectors were performed as an initial step in evaluating the feasibility of using an AAV vector for gene therapy. In this respect it was significant that we demonstrated complementation of the CF defect in cells derived from bronchial epithelium, since this is the site of the major clinical manifestation of the disease and is the most likely site for targeting of gene therapy vectors.

The observation that pSA306 successfully complemented the CF defect was especially pertinent. Previous studies had been interpreted as suggesting that the NH_2-terminal region of CFTR that is deleted in pSA306 was in fact essential for CFTR function when CFTR was expressed from various other vectors such as vaccinia (ANDERSON et al. 1991). The experiment with SA306 as a transducing vector showed that the efficiency of the vector allowed for expression of CFTR without selection in an airway cell population. This is analogous to the in vivo situation in which selection cannot readily be used. The work with AAV vectors also led to the novel observation that complementation of the IB3 cells with the AAV vectors resulted in both the appearance of the small (10 picosiemens) linear chloride conductance associated previously with recombinant CFTR expression and the return of normal cAMP regulation of the outwardly rectifying chloride channel associated previously with the CF defect in airway epithelia (EGAN et al. 1992). This was the first study which showed the relationship between these two channels. The relative contribution of either channel to the pathophysiology of CF is unknown. However, one implication is that AAV vectors may enhance the possibility of correction of the pathophysiology if it is indeed important that the function of both channels is correctly regulated.

Because AAV vectors represent a potentially important vector system for human gene therapy, it was important to solve the CFTR size problem. We did this because of the serendipitous observations with the SA306 construct. A possible alternate approach would be to use mutations which delete other regions of the CFTR protein but leave its function intact. One report suggested that a significant portion of the R domain can be deleted and still retain function, but in this case the regulation of the protein was altered (RICH et al. 1991). In this respect it may be noteworthy that the SA306 vector appears to be correctly regulated. However, for initial clinical trials, a vector containing the full CFTR cDNA is preferred because, as already noted, the pathophysiology of CF is not clearly understood. It is therefore difficult to predict if a truncated CFTR protein would be efficacious even if it appears to have correct electrophysiological function in vitro.

5 In Vivo Studies of Adeno-associated Virus-Cystic Fibrosis Transmembrane Regulator Vectors

To develop AAV vectors for in vivo delivery of a CFTR gene, a number of issues needed to be addressed. First, there were no data on in vivo delivery of any AAV vector. The efficiency of AAV vector delivery to the airway surface in vivo remained to be determined. The vector-host interactions also required investigation. It was unknown if immune or inflammatory responses would cause resistance to vector infection, adverse reactions for the host, such as bronchospasm or hypersensitivity pneumonitis or immune compromise of reinfection with subsequent doses of vector. The fate of integrated AAV vector genomes in intact airways remained a matter of speculation. While it was possible that long term expression might occur, it was also possible that vectors or vector particles might remain intact or might be rescuable and subsequently shed if wild-type AAV and helper virus infection occurred simultaneously. Shedding might be of concern if it occurred while the subject was in contact with other individuals. These initial questions needed to be carefully examined in appropriate animal models before AAV-based gene therapy in humans was considered.

We employed two in vivo animal models to begin to address some of these issues. It should be noted that once the initial demonstration of gene transfer and duration of expression is analysed several questions will remain to be addressed. For instance, it will be important to assess the need and possibility of repeated delivery of the vector and to dertermine if there are any humoral immune responses. Also, it is currently not known whether targeting a specific cell population in the airway will be important for reconstitution of CFTR function.

The first in vivo model (FLOTTE et al. 1993b) employed direct bronchoscopic delivery of an AAV CFTR transducing vector to a portion of one lobe of the lung in New Zealand White rabbits. This model was aimed at addressing issues of vector transduction or gene transfer efficiency, vector persistence, and CFTR gene expression as well as some initial evaluation of possible toxicities.

In a second in vivo model, we employed a similar bronchoscopic delivery of AAV CFTR vector to a portion of one lobe of the lung of rhesus macaques. This nonhuman primate model was mainly developed as a preclinical toxicology model and also to address some of the issues of vector mobilization and vector rescue.

In the rabbit in vivo model, we used the SA306 vector because the presence of the NH_2-terminal epitope tag provided a useful region to allow definitive evidence that the CFTR expression observed was indeed from the recombinant vector. In the rhesus macaque model, we used the vector tgAAVCF which has the entire CFTR cDNA expressed directly from the AAV ITR promoter. Since tgAAVCF is the vector designated for the proposed clinical tiral, it contains the entire CFTR cDNA and no additional foreign epitope tags. For the in vivo experiments we modified the vector production system to use the packaging plasmid pRS5 which, as noted above, gave a higher titer of vector particles (FLOTTE et al. 1995).

5.1 In Vivo Delivery of AAV-CFTR Vectors in Rabbits

5.1.1 Selective Bronchoscopic Delivery

The initial in vivo analysis of in vivo delivery of AAV-CFTR vectors to the lung was reported by FLOTTE et al. (1993b). Eleven adult New Zealand White rabbits were anesthesized with intramuscular ketamine and underwent flexible fiberoptic bronchoscopy with an Olympus 3.5-mm pediatric bronchoscope. The orifices of the major airways were visualized and the tip of the scope was wedged in the right lower lobe (RLL) bronchus. The SA306 AAV CFTR vector was then introduced into the RLL bronchial lumen through the suction port. In animals which served as negative controls, vehicle solution (excipient) was placed in a similar manner. Noninstrumented animals served as additional negative controls. Animals were sacrificed at 3 and 10 days and 3 months and 6 months postinfection and tissue samples were taken from both the targeted RLL and the left upper lobe (LUL). Both proximal and distal bronchi and trachea were sampled. Fixed sections were used for an in situ DNA PCR assay to detect presence of vector DNA and for immunohistochemistry to detect CFTR protein expression. Unfixed sections were used for detection of CFTR mRNA expression by RT-PCR and for immunodetection of the protein by western blot.

The immunodetection of CTFR protein was performed using two different chicken IgG antibodies (Flotte et al. 1993b). One primary antibody was a chicken anti-human CFTR R-domain antibody, number 602, and the other primary antibody, number 934, was a chicken polyclonal antibody directed against the 26 amino acid polypeptide epitope tag unique to the pSA306 vector (MLLIYVHTKNQHTLIDASELFIRPGT). This sequence bears no homology to native CFTR and so was chosen to eliminate any potential difficulties due to cross-reactivity with endogenous CFTR.

5.1.2 In Vivo DNA Transfer to Rabbit Airways

The presence of AAV CFTR vector DNA in rabbit lung sections was assessed by in situ DNA PCR (Flotte et al. 1993b). At either 3 or 10 days, vector DNA was detected in both the large and small airways and in the alveoli of the targeted lobe in each of the vector-treated animals. None was found in the nontargeted control lobe nor in any of the control animals. The lobar bronchus from the proximal RLL showed vector present in about 50% of cells including some cell nuclei near the basal lamina. In the distal airway the pattern was more patchy in that some smaller airways showed vector DNA present in nearly 100% of the cells, whereas others had few or no positive nuclei.

At 3 months postadministration, vector DNA was still detectable in about 50% of nuclei within the airway epithelium of the lobar bronchus of the RLL. By 6 months the proportion of cell with detectable vector DNA signal had decreased to about 5% and these positive cells occurred mostly in small patches.

Several other tissues were screened for recombinant vector DNA. In addition to the uninfected LUL, the proximal trachea and kidney were examined. Kidney sections were negative. Four of 16 tracheal sections examined had some detectable positive nuclei (about 5%–25% in positive sections) at each of the time points examined. This was most likely due to coughing or mucociliary transport of vector particles at or shortly after the time of vector administration. Hematoxylin/eosin-stained sections of lungs, heart, liver, and kidneys were examined. There were no differences between controls and experimental animals, even as late as 6 months after vector administration. All of the treated animals remained clinically healthy throughout the experiments. These findings suggested that in rabbits AAV CFTR was not toxic, either acutely or chronically.

5.1.3 Recombinant CFTR Protein Expression Parallels Vector DNA Distribution

The pattern of CFTR protein expression in the vector-treated animals was similar to that observed with vector DNA distribution. Both large and small airways and alveoli of the targeted lobe showed (Flotte et al. 1993b) increased immunoreactivity with the polyclonal anti-human CFTR-R domain antibody 602 and the anti-fusion epitope antibody 934 at 3 and 10 days after infection. Antibody 934 allowed the unambiguous distinction of vector-expressed protein from native CFTR, while antibody 602 directly confirmed that the CFTR polypetide sequence was expressed. Colocalization of CFTR immunoreactivity with antibodies 934 and 602 in serial sections provided additional evidence of the specificity of the immunodetection techniques.

In the large airways, the CFTR immunoreactivity was localized near the apical surface in many cells, which is analogous to the distribution pattern seen with endogenous CFTR. CFTR protein expression persisted in patches of cells in vector-treated lobes at 3 and 6 months. The controls, including the nontargeted LUL, the vehicle-instilled and the noninstrumented animals, all showed very low levels of immunoreactivity with antibody 602 or 934. The specificity of chicken anti-R domain antibody 602 was confirmed by immunoblotting and the specificity of the anti-fusion peptide antibody 904 was also demonstrated by immunoblotting (FLOTTE et al. 1993b). A protein with the expected apparent molecular weight of 170 000 was seen in protein samples from the vector-treated animals, but not from a vehicle-treated control animal.

5.1.4 Correlation of CFTR RNA and Protein Expression

To confirm that AAV CFTR vector DNA was acting by increasing recombinant CFTR protein levels directly and not by activating endogenous CFTR expression, we analyzed recombinant mRNA expression from tissue homogenates by reverse transcriptase-PCR (Fig. 4). Signal was detectable at 3 or 10 days in vector-treated animals, but not in the vehicle-treated controls. RNA transcribed from AAV CFTR was still detected 3 or 6 months after infection. This showed that CFTR mRNA was expressed from the vector for up to 6 months although the relative levels of mRNA expression could not be determined with the non-quantitative PCR assay.

5.1.5 Dose Response of In Vivo Delivery of AAV-CFTR in Rabbit Lung

Additional experiments showed a clear dose response between the amount of AAV CFTR vector delivered and the number of cells containing AAV CFTR vector genomes. There was also concordance in the dose response between the number of cells containing vector DNA and the number positive by immunocytochemistry (FLOTTE et al. 1994b).

5.1.6 Safety, Toxicity and Host Response

An additional observation in the rabbit experiments was that the targeted bronchoscopic delivery resulted in very little spillover to a nontargeted lobe, although there was detectable vector DNA in the trachea of some animals. No vector DNA was detectable in any other organs, confirming the selectivity of vector gene transfer. Most notably, there were no pathological changes sug-

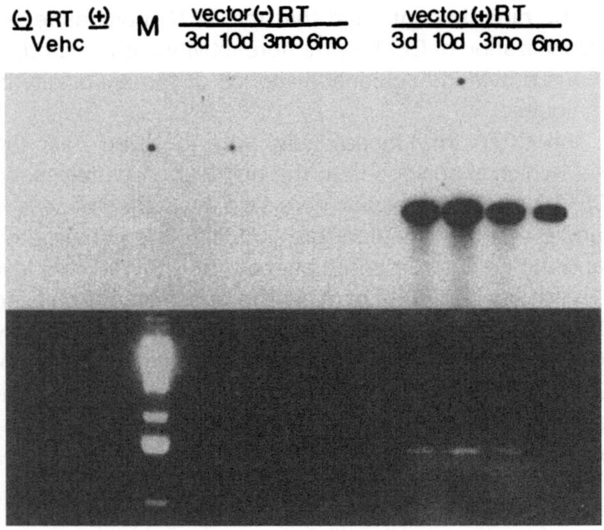

Fig. 4. Adeno-associated virus-cystic fibrosis transmembrane regulator RNA expression persists in rabbit lung for 6 months after infection. A Southern blot of products of a reverse transcriptase-PCR assay, performed on DNase-treated samples of total cellular RNA extracted from rabbit lung homogenates harvested 3 days, 10 days, 3 months, and 6 months after vector instillation, is shown. CFTR RNA expression at each of these time points in vector-treated animals is demonstrated by the amplified fragment. Control lanes include RNA from a lung homogenate from a vehicle-treated animal (Vehc) in which there is no signal. Duplicates of each sample handled as above but without reverse transcriptase, (–) RT lanes, demonstrate the completeness of vector DNA digestion. (From Flotte et al. 1993b)

gestive of inflammatory responses or neoplasia in the lungs or any other organs, suggesting that AAV CFTR vectors may be nontoxic and safe when delivered for airway gene therapy.

Another aspect of in vivo delivery which warranted consideration was the possibility that, in the course of vector delivery through a fiberoptic bronchoscope, the mucosal barrier might be breached resulting in significant systemic spread of recombinant virus. This possibility was studied by directly injecting a high dose (10^{10} particles) of AAV CFTR vector into the auricular veins of New Zealand white rabbits (T. Flotte, unpublished). Animals were killed at 1 week after vector injection and the spread of vector DNA was studied by standard DNA PCR and by in situ DNA PCR of various organs including RLL lung, LUL lung, trachea, thymus, bronchial lymph node, heart, liver, spleen, pancreas, kidney, jejunum, mesenteric lymph node, gonad, and brain. No vector DNA was detectable in a vehicle-injected control animal, but vector DNA was present in the thymus of both of the vector-injected animals, and in the bronchial lymph node of one of two vector-injected animals. The finding of spread to lymphoid tissues in consistent with previous studies showing AAV infection of lymphocytes (Mendelson et al. 1992). No vector DNA was detectable in the gonads or any other organs of either of the vector-injected animals. This study indicated that while hematogenous dissemination of AAV CFTR may occur in rabbits after direct intravenous injection, this is not likely to result in germ-line alteration, even in the unlikely event that an entire bolus of the highest vector dose were to enter the bloodstream. Furthermore, there was no evidence of pathologic changes in any organ as a result of intravenous injection of vector.

5.1.7 Summary of In Vivo Delivery to Rabbit Lung

The rabbit experimental model demonstrated the efficacy of AAV vectors for in vivo gene transfer to the airway. The presence and expression of vector genomes in cells from the airway surface epithelium were seen as early as 3 days after infection and persisted for at least 6 months. This suggested that AAV vectors may be capable of long-term expression after in vivo gene transfer to the airway epithelium which would be an important advantage for CF gene therapy.

The "epitope-tagging" of CFTR with a novel fusion peptide sequence allowed the design of vector-specific antibodies and probes. These reagents facilitated the distinction of vector RNA and protein expression from endogenous CFTR expression to establish biological efficacy in the current model. These studies also demonstrated the ability to physically target a gene therapy vector to a single lobe of the lung by fiberoptic bronchoscopy. This rabbit bronchoscopic model provided a way to infect cells with the multiplicity of virus particles needed to infect a large percentage of cells in the airway, in the absence of a selective marker.

Several issues remain to be addressed. The long duration of expression observed in vivo might be due to AAV vector integration but this remains to be demonstrated. Furthermore, initial evidence from the rhesus monkey studies suggests that it is possible that AAV vectors may persist in vivo as episomes. Since the lifespan of mammalian airway epithelial cells may be about 120 days (ROSENFELD et al. 1991), it is possible that in the rabbit model the decrease in the number of cells with vector DNA and recombinant protein signal by 6 months was due to loss of the more terminally differentiated surface epithelium, with persistence of patches of cells derived from a stably transduced precursor cell population. The identity of precursor stem cells remains to be clearly understood but in the rabbit these may be represented by basal cells. In this respect, it is of significance that the AAV vector could be detected in a proportion of the basal cells (C. Conrad and T. Flotte, unpublished).

5.2 In Vivo Delivery of AAV-CFTR Vectors in Rhesus Macaques

Rhesus monkeys (FLOTTE et al. 1994b) were chosen for the final preclinical phase of AAV CFTR vector testing since these primates provide a suitable host for AAV and adenoviruses. They are particularly relevant in the context of identifying potential problems related to recombinant virus shedding, immunologic reactions, or other potential adverse effects from AAV CFTR vectors delivered to the airway surface via a flexible fiberoptic bronchoscope.

5.2.1 Bronchoscopic Delivery Rhesus Airways

This study of AAV CFTR vector safety and biological activity was performed in 3 kg young adult rhesus macaques. These animals represent a model species closer to humans and more suitable for a variety of preclinical safety assessments which more closely parallel a clinical protocol. A variety of vector doses and time points were studied, along with vehicle controls. Animals were sedated with intramuscular ketamine and underwent halothane anesthesia. Baseline studies were performed including passive measurements of pulmonary mechanics with a Sensormedics 2600 infant pulmonary function test (PFT) machine, chest X-rays, and arterial blood gases. Next, a 3.5 mm pediatric bronchoscope was passed through an oral airway and into the RLL bronchus. After bronchoalveolar lavage (BAL), to evaluate cell counts and the levels of interleukin (IL)-6 and -8, doses of vector were instilled. PFTs were then repeated and finally animals were killed at 10 days, 21 days, 90 days, or 180 days.

Prior to sacrifice, pulmonary function tests, chest X-rays, arterial blood gases and BAL were repeated. After killing the animals, various organs including lungs (all five lobes), trachea, thymus, heart, bronchial lymph nodes, liver, spleen, pancreas, kidney, jejunum, mesenteric lymph node, brains and gonads were sampled.

Separate samples of each organ were either fixed in 10% buffered formalin for histopathologic assessment (after embedding, sectioning and H&E staining) and for in situ DNA PCR (with vector-specific primers) or processed for DNA and RNA isolation for use in DNA PCR and RT-PCR assays, respectively. Additionally, cells were harvested from the RLL bronchus for primary culture and further molecular characterization.

The primary endpoints were: (1) DNA PCR and in situ DNA PCR to survey for the presence of AAV CFTR vector in multiple sites, (2) RT-PCR to assess vector expression, (3) pulmonary function tests to assess for any immediate airway reactivity to the vector, (4) arterial blood gas measurements to assess any ventilation-perfusion inequality, (5) chest X-ray and histopathology to visualize any infiltrative alveolar processes, and (6) BAL cell counts to assess potential airway or alveolar inflammatory responses.

5.2.2 Vector DNA Distribution and Dose Response

The principal purposes of this rhesus toxicology study were to confirm that AAV CFTR gene transfer and expression had occurred without significant toxicity. In each of the vector-treated animals, vector DNA presence and expression were detectable in the vector-treated RLL. Vector DNA was not detectable in cells from the RLL of the control animals. At the highest dose, a low level of vector

DNA was also present in the LUL of the lung and the trachea of vector treated animals. The proportion of cells in the RLL which contained vector (as measured by in situ PCR) increased in a dose-dependent fashion. At the highest doses, more than 50% of the cells contained vector. For a given dose, the proportion of cells containing vector was constant over the period of 10–180 days post–instillation.

The presence and expression at very low levels of AAV CFTR vector DNA were also detected in the liver of some animals, and in the trachea. However, there were no significant abnormalities noted in any of the safety assessments. The functional studies of airway function and gas exchange (arterial blood gases, chest radiographs) were entirely normal. Although spread to the liver was observed by in situ PCR it was localized to one small area (<1% of the total area of one cross-section), and there was no spread to the gonads, even at this very high dose, and no evidence of neoplastic changes. Taken together these studies indicate that AAV CFTR vector administration is safe in primates at the doses tested, which ranged up to 1×10^{11} particles administered to a single lung lobe and at time points ranging from 10 days to 180 days after vector instillation.

5.2.3 Safety, Toxicity and Host Response

Although, in the monkey model, there was some spread to some other organs following pulmonary delivery, histopathological examination of those tissues showed no abnormalities. Furthermore, there was no indication of either short term or long term toxicities. There was no indication of any cellular immunological response or infiltration by components of the cellular immune response. There were no obvious inflammatory responses and there were no vector-related fluctuations in the inflammatory cytokines IL-6 or IL-8.

It is also worth noting that the monkeys used in these studies were seropositive for AAV as measured in an ELISA assay prior to vector instillation. Following administration of vector, there were no vector-related changes in the serum levels of AAV antibody. It is possible that this may have represented prior exposure of the monkeys to AAV4 which can react with the AAV2 serotype used for the vector. Nevertheless, this suggests that AAV seropositivity does not preclude function of the vector. This is important since most adult CF patients in clinical trials are likely to be seropositive for AAV which is reflective of the AAV seropositivity of most adults.

5.2.4 Vector Mobilization and Shedding

One additional safety issue that was studied in rhesus macaques was the potential for spread of recombinant AAV CFTR from vector-treated individuals who might subsequently be infected with wild-type AAV2 and adenovirus. In cultured cells transduced with AAV CFTR recombinants, infectious vector subsequently can be "rescued" when the cells are infected with wild-type AAV and helper adenovirus. In order to determine whether this phenomenon would occur in the complex context of an in vivo infection, we studied two rhesus monkeys that were infected with AAV CFTR vector in the RLL (Afione et al. 1994). Ten weeks later these animals were infected intranasally with a large dose of wild-type AAV2 and an adenovirus type 2 host-range mutant (Ad2HR405), which was known to possess AAV2 helper function and to be able to infect monkey cells in culture. Viral cultures were performed prior to infection and at 3 days, 7 days, 14 days, and 21 days from BAL fluid and nasal washings. These cultures were performed in 293 cells under three conditions: (1) inoculation of the specimen alone, to detect any replicating adenovirus, (2) inoculation of the specimen together with adenovirus-5 in order to detect replicating AAV2 and (3) inoculation of the specimen with adenovirus-5 and AAV2 to detect infectious recombinant AAV CFTR. After 2 days, analysis of DNA was performed with both AAV2 and CFTR specific probes to identify wild-type and recombinant AAV DNA replicating forms.

The results of these studies were as follows. Ad2HR405 infection was clearly established in the monkeys since adenovirus replication was detectable (by cytopathic effect and the presence of infectious adenovirus in both BAL and nasal wash cultures from the 3 day and 7 day samples, while none was seen in the pretreatment fluids. AAV2 replication was also established in one of the two monkeys at 3 days, but only in the nose. No recombinant AAV CFTR was detectable in any fluid at any time point.

These data indicate that while rhesus monkeys can have experimental adenovirus and AAV infections established, the AAV infection remains localized to the upper respiratory tract, near its normal portal of entry, and does not appear to spread significantly to the lower tract. Since AAV replication did not occur in the lower tract where vector was localized, there appeared to be little opportunity for in vivo rescue of the vector to occur.

One alternative explanation was that AAV CFTR vectors are not present in a rescuable form in the bronchial epithelial cells of these animals. However, we obtained direct evidence that primary bronchial epithelial cells grown from monkeys infected in vivo can undergo AAV CFTR vector rescue in vitro. In these studies, primary RLL bronchial epithelial cells from a macaque 3 months after the instillation of a 10^{11} dose of AAV-CFTR vector were infected with adenovirus-5 alone or adenovirus-5 and AAV2. Hirt supernatants were examined by Southern blotting with AAV and CFTR probes. As expected, AAV2 replicating DNA was detected in the $Ad5^{+}/AAV^{+}$ infected cells. A large amount of recombinant CFTR vector DNA was also found in these cells as evidenced by the presence of CFTR-hybridizing DNA species in the $Ad5^{+}/AAV^{+}$ infected cells. These vector DNA species included bands of the expected sizes for monomeric and dimeric replicating form DNA, indicating that intact copies of vector DNA wer indeed present in rescuable forms in these cells. This showed that the vector genome can persist in vivo in the lung in an intact form for at least 3 months. More recent experiments (T. Flotte and S. Afione, unpublished) suggest that the vector may persist at least in part as an episome whereas integration of the vector at a chromosomal site in vivo has not yet been shown definitively.

5.2.5 Summary of In Vivo Delivery to Rhesus Lung

Taken together, the results of the in vitro and in vivo rescue experiments indicate that although AAV CFTR vectors are generally present in rescuable form in vector-treated animals, vector rescue does not frequently occur following in vivo wild-type infection. It also appears that it is the regional distribution of wild-type AAV2 infection in the upper respiratory tract which makes in vivo rescue unlikely to occur. The studies also suggest that AAV vectors can persist and express in primate lungs for up to 6 months in the absence of significant toxicity. Even though there was some spread to other organs, this also did not cause any obvious toxicities.

6 Human Clinical Trials of Adeno-associated Virus-Cystic Fibrosis Transmembrane Regulator Vectors

Based on the preclinical studies in rabbits and rhesus macaques, a phase I clinical trial of direct in vivo delivery of AAV CFTR vectors was recently approved by the NIH Recombinant Advisory Committee (Flotte et al. 1994a).

This will be a single center, phase I, open-label, nonrandomized, dose escalation study in adult CF patients and mild to moderate lung disease. Each individual will receive a single dose of AAV CFTR vector administered to the nasal epithelium alone or to the nasal epithelium and the RLL of the lung.

Patients will be studied on the day of admission (day–4) to provide baseline data. Fiberoptic bronchoscopy will be performed on the day of admission for the completion of the baseline studies (CFTR RNA and protein assays on brushed cells and IL-6 and IL-8 levels in BAL fluid) and for the administration of vehicle to the superior segment of the RLL. The following 4 days will serve as the vehicle control period. If the patient's viral cultures from the nose, blood, urine, and stool are negative for infectious adenovirus and infectious AAV, the patient will then proceed into the vector administration phase of the study. Intranasal doses of vector or vehicle will be given to the surfaces of the right and left inferior turbinates by catheter. The investigator will be blinded as to the assignment of vehicle and vector to individual nostrils to provide for nonbiased nasal potential difference measurements. Endobronchial doses will be administered to the superior segment of the RLL on the day following nasal instillation. The nasal instillation may provide the best opportunity to detect vector activity due to the ease of access for cell samples and the ability to perform transmembrane potential difference measurements. However, in other clinical trials of CF gene therapy, it has begun to be recognized that this measurements is difficult and may be subject to significant variability (Alton and Geddes 1995; Caplen et al. 1995). The endobronchial instillation will provide the most relevant toxicity data.

Studies will be performed at 3 days and 10 days postvector instillation. If viral cultures taken at day 10 are negative, the patient will be discharged without any further interventions or precautions. If cultures are positive from nose, urine or stool for recombinant AAV CFTR vector at this time point, the patients will be instructed as to proper home precautions and decontamination procedures. AAV is generally transmitted via particles rather than aerosols and is easily decontaminated with ethanol or bleach. The likelihood of this circumstances arising is expected to be quite low. In any case, the environmental risk derived from the possible spread of AAV CFTR vectors is likely to be minimal, particularly since both wild-type AAV and concurrent adenovirus functions are necessary for active replication.

Acknowledgements. We thank the many individuals who have contributed to the studies described here including Sandra Afione, Carol Conrad, Rikki Solow, Bill Guggino, Pam Zeitlin and Marie Egan at Johns Hopkins Hospital and Tom Reynolds, Molly FeeMaki, and Caroleen Becker at Targeted Genetics Corporation.

References

Afione S, Conrad C, Adams R, Carter B, Flotte T (1994) AAV-CFTR vector rescue in Rhesus monkeys. Pediatr Pulmonol [Suppl] 10: 220

Alton EWFW, Geddes DM (1995) Gene therapy for cystic fibrosis: a clinical perspective. Gene Ther 2: 88–95

Anderson MP, Gregory RJ, Thompson S, Souza DW, Mulligan R, Smith AE, Welsh MJ (1991) Demonstration that CFTR is a chloride channel by alteration of its anion selectivity. Science 253: 202–205

Barasch J, Kiss B, Prince A, Saiman L, Gruenert D, Al-Aqwati Q (1991) Defective acidification of intracellular organelles in cystic fibrosis. Nature 352: 70–73

Bear CE, Duguay F, Naismith AL, Kartner N, Hanrahan JW, Riordan JR (1991) Cl^- channel activity in Xenopus oocytes expressing the cystic fibrosis gene. J Biol Chem 266: 19142–19145

Beaton A, Palumbo P, Berns KI (1989) Expression from the adeno-associated virus p5 and p19 promoters is negatively regulated in *trans* by the *rep* protein. J Virol 63: 4450–4454

Blacklow NR (1985) Adeno-associated viruses of humans. In: Pattison J (ed) Parvoviruses and human disease. CRC Press, Boca Raton, pp 165–174

Boat T, Welsh MJ, Beaudet A (1989) Cystic fibrosis. In: Scriver C, Beaudet A, Sly W, Valle D (eds) The metabolis basis of inherited disease. McGraw-Hill, New York, pp 2649–2860

Briggs MR, Kadonga JT, Bell SP, Tjian R (1986) Purification and biochemical characterization of the promoter-specific transcription factor Sp1. Science 24: 47–52

Caplen NJ, Alton EWFW, Middleton PG, Dorin JR, Stevenson BJ, Gao X, Durham SR, Jeffery PK, Hodson ME, Coutelle C, Huang L, Porteous DJ, Williamson R, Geddes DM (1995) Liposome-mediated CFTR gene transfer to the nasal epithelium of patients with cystic fibrosis. Nature Med 1: 39–46

Carter BJ (1990) The growth cycle of adeno-associated virus. In: Tjissen P (ed) Handbook of parvoviruses, (vol 1) CRC Press, Boca Raton, pp 155–168

Carter BJ (1992) Adeno-associated virus vectors. Curr Opin Biotechnol 3: 533–539

Carter BJ, Mendelson E, Trempe JP (1990) AAV DNA replication, integration and genetics. In: Tjissen P (ed) Handbook of parvoviruses, (vol 1) CRC Press, Boca Raton, pp 169–226

Chen JH, Schulman H, Gardner P (1989) A cAMP-regulated chloride channel in lymphocytes that is affected in cystic fibrosis. Science 243: 657–660

Crystal RG, McElvaney NG (1995) IL6 release and airway administration of human CFTR cDNA adenovirus vector. Nature Med 1: 182–184

Crystal RG, McElvaney NG, Rosenfeld MA, Ch C-S, Mastrangeli A, Hay JG, Brody SL, Jaffe HA, Eissa NT, Danel C (1994) Administration of an adenovirus containing the human CFTR cDNA to the respiratory tract of individuals with cystic fibrosis. Nature Genet 8: 42051–41057

DePamphilis ML (1988) Transcriptional elements as components of eukaryotic origins of DNA replication. Cell 52: 635–638

Drumm ML, Pope HA, Cliff WH, Rommens JM, Marvin SA, Tsui L-C, Collins FS, Frizzell RA, Wilson JM (1990) Correction of the cystic fibrosis defect in vitro by retrovirus-mediated gene transfer. Cell 62: 1227–1233

Egan M, Flotte T, Zeitlin P, Afione S, Solow R, Carter BJ, Guggino WB (1992) Correction of defective PKA regulation of outwardly rectifying chloride channels by insertion of normal CFTR into airway cells. Nature 359: 581–584

Flotte T, Owens RA, Afione SA, Solow R, Zeitlin PL, Carter BJ (1992) Gene expression from adeno-associated virus vectors in airway epithelial cells. Am J Respir Cell Mol Biol 7: 349–356

Flotte TR, Zeitlin PL, Solow R, Afione S, Owens RA, Markakis D, Drumm M, Guggino WB, Carter BJ (1993a) Expression of the cystic fibrosis transmembrane conductance regulator from a novel adeno-associated virus promoter. J Biol Chem 268: 781–3790

Flotte TR, Afione SA, Solow R, McGrath SA, Conrad C, Zeitlin PL, Guggino WB, Carter BJ (1993b) In vivo delivery of adeno-associated vectors expressing the cystic fibrosis transmembrane conductance regulator to the airway epithelium. Proc Natl Acad Sci USA 93: 10163–10617

Flotte T, Carter B, Conrad C, Guggino W, Reynolds T, Rosenstein B, Taylor G, Walden S, Wetzel R (1994a) A phase I study of an adeno-associated virus CFTR gene vector in adult cystic patients with mild lung disease. NIH Recombinant Advisory Committee, Bethesda, Maryland, 12 Sept

Flotte TR, Conrad CK, Afione SA, Reynolds TC, Guggino WB, Carter BJ (1994b) Preclinical trials of adeno-associated virus (AAV) vectors for cystic fibrosis (CF) gene therapy. Pediatr Pulmonol [Suppl] 10: 153–154

Flotte TR, Barraza-Ortiz X, Solow R, Afione SA, Carter BJ, Guggino WB (1995) An improved system for packaging recombinant adeno-associated virus vectors capable of in vivo transduction. Gene Ther 2: 1–9

Grossman Z, Mandelson E, Brok-Simoni F, Mileguir F, Leitner Y, Rechavi G, Ramot B (1992) Detection of adeno-associated virus type 2 in human peripheral blood. J Gen Virol 73: 961–966

Hai T, Liu F, Allegretto EA, Karin M, Green MR (1988) A family of immunologically related transcription factors that includes multiple forms of ATF and AP-1. Genes Dev 2: 1216–1226

Hermonat PL, Muzyczka N (1984) Use of adeno-associated virus as a mammalian cDNA cloning vector: transduction of neomycin resistance into mammalian tissue culture cells. Proc Natl Acad Sci USA 81: 6466–6470

Hwang T-C, Lu L, Zeitlin PL, Gruenert DC, Huganir R, Guggino WB (1989) Chloride channels in CF: lack of activation by protein kinase C and cAMP-dependent protein kinase. Science 244: 1352–1353

Kaplitt MG, Leone P, Samulski RJ, Xiao X, Pfaff DW, O'Malley KL, During MJ (1994) Long-term expression and phenotypic correction using adeno-associated virus vectors in the mammalian brain. Nature Genet 8: 148–154

Kartner N, Hanarahan JW, Jensen TJ, Naismaith AL, Sun S, Ackerley CA, Reyes EF, Tsui LC, Rommens LM, Bear CE, Riordan JR (1991) Expression of the cystic fibrosis gene in nonepithelial invertebrate cells produces a regulated anion conductance. Cell 64: 681–691

Kotin RM, Berns KI (1989) Organization of adeno-associated virus DNA in latently infected Detroit 6 cells. Virology 170: 460–467

Kotin RM, Siniscalo M, Samulski RJ, Zhu X, Hunter L, Laughlin CA, Muzyczka N, Rocchi M, Berns KI (1990) Site-specific integration by adeno-associated virus. Proc Natl Acad Sci USA 87: 2211–2215

Laughlin A, Tratschin JD, Coon H, Carter BJ (1983) Cloning of infectious adeno-associated virus genomes in bacterial plasmids. Gene 23: 65–73

Lebkowski JS, McNally MM, Okarma TB, Lerch LB (1988) Adeno-associated virus: a vector system for efficient introduction of DNA into a variety of mammalian cell types. Mol Cell Biol 8: 3988–3994

Li M, McCann JD, Liedtke CM, Nairn AC, Greengard P, Welsh MJ (1988) Cyclic AMP dependent-protein kinase opens chloride channels in normal but not in cyctic fibrosis airweay epithelium. Nature 331: 358–360

Lin YS, Green MR (1988) Interaction of a common transcription regulatory factor, ATF, with regulatory elements in both E1A- and cyclic Amp-inducible promoters. Prioc Natl Acad Sci USA 85: 3396–3340

Luo L, Zeitlin PL, Guggino WB, Craig RW (1989) Gene transfection by lipofection in rabbit and human secretory epithelial cells. Pflugers Arch 415: 198–203

McLaughlin SK, Collis P, Hermonat PL, Muzyczka N (1988) Adeno-associated virus generalized transduction vectors: analysis of proviral structures. J Virol 62: 1963–1973

Mendelson E, Grossman Z, Mileguir F, Rechavi G, Carter BJ (1992) Replication of adeno-associated virus type 2 in human lymphocytic cells and interaction with HIV-1. Virology 187: 453–563

Miller DG, Adam MA, Miller AD (1990) Gene transfer by retrovirus vectors occurs only in cells that are actively replicating at the time of infection. Mol Cell Biol 10: 4239–4242

Mitchell PJ, Tjian R (1989) Transcriptional regulation in mammalian cells by sequence-specific DNA binding proteins. Science 245: 371–378
Montminy M, Bilezkjian LM (1987) Binding of a nuclear protein to the cyclic-AMP response element of the somastatin gene. Nature 328: 175–178
Montminy M, Gonzalez GA, Yamamoto KK (1990) Regulation of cAMP-inducible genes by CREB. TINS 13: 184–188
Muzyczka N (1992) Use of Adeno-associated virus as a generalized transduction vector in mammalian cells. In: Muzyczka N (ed) Viral expression vectors. Springer, Berlin Heidelberg New York, pp 97–129 (Current topics in microbiology and immunology, vol 158)
Pitluk ZW, Ward DC (1991) Unusual Sp1-GC box interaction in a parvovirus promoter. J Virol 65: 6661–6670
Rich DP, Anderson MP, Gregory RJ, Cheng SH, Paul S, Jefferson DM, McCann DM, Klinger KW, Smith AE, Welsh MJ (1990) Nature 347: 358–363
Rich DP, Gregory RJ, Anderson MP, Manalavan P, Smith AE, Welsh MJ (1991) Effect of deleting the R-domain on CFTR-generated chloride channels. Science 253: 205–207
Riordan JR, Rommens JM, Kerem N, Alon N, Rozmahel R, Grzelczak Z, Lok S, Plavsic N, Chou J-L, Drumm ML, Iannuzzi MC, Collins FC, Tsui L-C (1989) Identification of the cystic fibrosis gene: cloning and characterization of complementary DNA. Science 245: 1066–1073
Rosenfeld MA, Yoshimura K, Trapnell BC, Yoneyama K, Rosenthal ER, Dalemans W, Fukayama M, Stratford-Perricaudet L, Perricaudet M, Guggino WB, Pavirani A, Lecoq JP, Crystal RG (1991) In vivo transfer of the human cystic fibrosis transmembrane conductance regulator gene to the airway epithelium. Cell 68: 143–155
Samulski RJ (1993) Adeno-associated virus integration at a specific chromosomal locus. Curr Opin Genet Dev 3: 74–80
Samulski RJ, Berns KI, Tan M, Muzyczka N (1982) Cloning of adeno-associated virus into pBR322: rescue of intact virus from the recombinant plasmid in human cells. Proc Natl Acad Sci USA 79: 2077–2081
Samulski RJ, Chang LS, Shenk TE (1989) Helper-free stocks of recombinant adeno-associated viruses: normal integration does not require viral gene expression. J Virol 63: 3822–3828
Samulski RJ, Zhu X, Xia X, Brook JD, Housman DE, Epstein N, Hunter LA (1991) Targeted integration of adeno-associated virus (AAV) into human chromosome 19. EMBO J 10: 3941–3950
Sassoni-Corsi P (1988) Cyclic AMP induction of adenovirus promoters involves sequences required for E1A trans-activation. Proc Natl Acad Sci USA 85: 7192–7196
Schoumacher RA, Shoemaker RL, Halm DR, Tallant EA, Wallace RW, Frizzell RA (1987) Phosphorylation fails to activate chloride channels from cystic fibrosis airway cells. Nature 330: 752–754
Schwiebert EM, Egan ME, Hwang TH, Fulmer SB, Allen SS, Cutting GR, Guggino WB (1995) CFTR regulates outwardly rectifying chloride channels through an autocrine mechanism involving ATP. Cell 81: 1063–1073
Smale ST, Baltimore D (1989) The "initiator" as a transcription control element. Cell 57: 103–113
Smale ST, Schmidt MC, Berk AJ, Baltimore D (1990) Transcriptional activation by Sp1 as directed through TATA or initiator: specific requirement for mammalian transcription factor IID. Proc Natl Acad Sci USA 87: 4509–4513
Tratschin J, Miller IL, Smith MG, Carter BJ (1985) Adeno-associated virus vectors for high frequency integration, expression and rescue of genes in mammalian cells. Mol Cell Biol 5: 3251–3260
Tratschin JD, West MHP, Sandbank T, Carter BJ (1984) A human parvovirus, adeno-associated virus as a eukaryotic vector: transient expression and encapsidation of the prokaryotic gene for chloramphenicol acetyl transferase. Mol Cell Biol 4: 2072–2081
Walsh CE, Liu JM, Xiao X, Young NS, Nienhius AW, Samulski RJ (1992) Regulated high level expression of a human gamma-globin gene introduced into erythroid cells by a novel adeno-associated virus vector. Proc Natl Acad Sci USA 89: 7257–7261
Ward CL, Krouse ME, Gruenert DC, Kopito RR, Wine JJ (1991) Cystic fibrosis gene expression is not correlated with rectifying Cl^- channels. Proc Natl Acad Sci USA 88: 5277–5281
Zabner J, Couture LA, Gregory RJ, Graham SM, Smith AE, Welsh MJ (1993) Adenovirus-mediated gene transfer transiently corrects the chloride transport defect in nasal epithelia of patients with cystic fibrosis. Cell 75: 207–216
Zeitlin PL, Lu I, Hwang TC, Rhim J, Cutting G, Stetten G, Kieffer KA, Craig R, Guggino WB (1991) A cystic fibrosis bronchial epithelial cell line: immortalization by adeno12-SV40 infection. Am J Respir Cell Mol Biol 4: 313–319

Adeno-associated Virus Based Vectors As Antivirals

K.K. Wong, Jr.[1] and S. Chatterjee[2]

1 Introduction

Although great strides have been made in the treatment and prevention of human viral infections, antiviral drug therapy still remains problematic when compared to treatment of bacterial infections. The therapeutic toxicity/efficacy ratio of many antivirals is low, resulting in potential risks of regimen related toxicity.

[1]Departments of Hematology and Bone Marrow Transplantation, Infectious Diseases, 1500 E. Duarte Road, Duarte, CA 91010, USA
[2]Division of Pediatrics, Beckman Research Institute, City of Hope National Medical Center, 1500 E. Duarte Road, Duarte, CA 91010, USA

The number of effective antivirals is limited, and many important viral infections remain untreatable. In addition, currently used antivirals often suppress rather than cure infections, necessitating repeated, sometimes lifelong, therapy to prevent recurrences. Furthermore, the increasing use of antivirals has resulted in the selection or emergence of drug-resistant strains as has occurred following the widespread use of antibacterials (FIELD and BIRON 1994). Finally, several medically important viruses, most notably human immunodeficiency virus (HIV), have been recalcitrant to either the development of curative chemotherapy or protective vaccines (SCHNITTMAN and FAUCI 1994). Thus, novel approaches to the treatment or prevention of viral infections are constantly being sought.

Recent advances in molecular biology and genetic engineering have fostered gene transfer strategies designed to augment antiviral immune responses or confer actual cellular resistance to specific viruses. The utility of altering cellular resistance to infection, termed "intracellular immunization" by BALTIMORE (1988), lies in the direct application of knowledge about molecular pathogenesis to precisely interrupt productive viral infection. Such strategies may be particularly useful in the management of intracellular pathogens which are capable of evading the immune system, or in circumstances in which knowledge of pathogenic mechanisms at the molecular level surpasses the ability to develop appropriate therapeutic drugs. Progress in this area has been most notable in agriculture, in which plants have been genetically modified to enhance resistance to a variety of pathogens (KANIEWSKI and THOMAS 1993). Similarly, transgenic animals have been derived which resist challenge by specific pathogens (SALTER and CRITTENDEN 1989; ARNHEITER et al. 1990; HAN et al. 1991). Perhaps not surprisingly, gene transfer approaches to the management of AIDS have become major areas of interest, and several human clinical trials of marrow-derived immune cells genetically modified to resist HIV infection are about to be initiated (NABEL et al. 1994; GALPIN et al. 1994). In this chapter, we briefly review current strategies for the gene therapy of viral infections with a focus upon the potential role of adeno-associated virus (AAV)-based vectors for the delivery of viral inhibitory molecules.

2 General Principles

Antiviral gene therapy requires stable expression of transgene(s) encoding effective, noncytotoxic antiviral molecules coupled with an efficient, nontoxic gene delivery system. In addition, conferring stable viral resistance to stem cells, thus producing a self-renewing, perpetually resistant cell population, is highly desirable. Although there is currently no consensus concerning the most effective antiviral targets, inhibition of *early* viral functions, at a time when viral gene products are least abundant and perhaps most easily inhibited, may be desirable, thereby preventing the accumulation of late viral products which may be deleterious to the cell. Additionally, inhibition of essential, highly conserved functions may afford cellular protection from serotypically diverse but genetically related viruses. Thus, production of both herpes simplex virus types 1 and 2 (HSV-1 and HSV-2) is significantly inhibited in cells constitutively expressing an antisense RNA complementary to the 5′-noncoding region of the HSV-1 ICP4 transcript (WONG et al. 1991). Similarly, cells expressing an antisense transcript complementary to both the conserved HIV-1 trans-activating response element (TAR) region and polyadenylation sites are protected from challenge from both HIV-1 and simian immunodeficiency virus (SIV) (CHATTERJEE et al.1992a). Fortunately, mutational analyses have identified a multitude of requisite virus-encoded genes which may constitute suitable targets, including the *tat* and *rev* genes of HIV, the *tax* gene of human T-cell leukemia virus (HTLV-1), the Ela gene of adenovirus, the VP16, and ICP4 genes of HSV, and the E6 and E7 transforming genes of human papillomaviruses (HPV) (reviewed in WONG and CHATTERJEE 1992).

3 Gene Delivery Systems: Eukaryotic Viral Vectors

Viral vectors have been derived to exploit the natural ability of viruses to efficiently infect and transfer their genetic material into cells. Although a wide variety of viruses are capable of infecting cells, few are both noncytopathic and stably integrate into cellular DNA. Vectors based upon murine retroviruses have been widely used for gene transfer to cells in vitro and in animal models in vivo (MILLER 1992a). In addition, these vectors have recently been used in human clinical trials for insertion of the wild-type adenosine deaminase (ADA) gene into ADA deficient hematopoietic cells, for induction of antiviral and antitumor immunity, and for genetic "marking" of hematopoietic stem cell transplants to determine sites of tumor relapse and patterns of hematopoietic stem cell reconstitution (BLAESE 1993; BRENNER et al. 1993; DUNBAR et al. 1995; for reviews see ANDERSON 1992; MILLER 1992b; MULLIGAN 1993). The life cycle of these viruses involves reverse transcription of genomic viral RNA and subsequent proviral integration into host cellular DNA. While retroviruses do not appear to integrate in a totally random fashion, proviral integration occurs in a relatively large number of sites and has been associated with both insertional mutagenesis (HARTUNG et al. 1986) and tumorigenesis due to insertion of viral transcriptional enhancer elements upstream of endogenous cellular oncogenes (CUYPERS et al.1984; HAYWARD et al. 1981; KUNG et al. 1991). Furthermore, replication competent retroviruses (RCRs), which may be generated during the encapsidation process and contaminate vector stocks, have been causally linked to the development of rapidly progressive T cell lymphomas in a primate model of transplantation with retroviral vector transduced progenitors (DONAHUE et al. 1992). Finally, successful proviral integration is highly dependent upon transduction of actively cycling cells

(MILLER et al. 1990; ROE et al.1993). This represents a formidable problem, as several important cellular populations including neurons and stem cells, attractive targets for genetic modification, are believed to be either quiescent or to cycle infrequently. These limitations are reflected in retroviral marking efficiencies of hematopoietic stem cells in vivo of < 5% in large animal models and < 1% in actual human clinical trials (KANTOFF et al. 1987; CARTER et al. 1992; BRENNER et al. 1993; DUNBAR et al. 1995).

Recently, eukaryotic vectors based upon the nonpathogenic parvovirus, AAV, have been proposed for antiviral gene therapy targeting HSV (WONG et al.1991), HIV (CHATTERJEE et al. 1992a), and HPV (LU et al. 1994a; SHILLITOE et al. 1994). AAV are replication defective DNA viruses which require defined helper functions provided by coinfection with another DNA virus, typically either adenovirus or a herpes virus, for productive infection (reviewed in BERNS and BOHENZKY 1987). As outlined here and elsewhere in this volume, AAV-based vectors offer a number of advantages for this purpose, including high transduction frequencies in cells from diverse species and lineages including hematopoietic progenitors (HERMONAT and MUZYCZKA 1984; TRATSCHIN et al. 1984; LAFACE et al. 1988; LEBKOWSKI et al. 1988; MCLAUGHLIN et al. 1988; reviewed in MUZYCZKA 1992; CHATTERJEE and WONG 1993; KOTIN 1994 and elsewhere in this volume); lack of superinfection inhibition, permitting multiple transductions with the same or different vector(s) (LEBKOWSKI et al. 1988); and integration into cellular DNA. Latent wild-type AAV infections have been maintained in tissue culture for greater than 100 serial passages in the *absence* of selective pressure (BERNS et al. 1975), attesting to the stability of this integration. Although wild-type AAV selectively integrates into a region of human chromosome 19 (19q13.3-qter) (KOTIN et al. 1990), this property may not extend to *rep*-deleted AAV vectors (Chatterjee et al., unpublished). Additionally, AAV vectors frequently integrate as multicopy tandem repeats (MCLAUGHLIN et al. 1988), potentially enhancing transgene expression. Pretreatment of cells with inhibitors of proliferation including aphidicolin (PODSAKOFF et al. 1994a), FUdR (PODSAKOFF et al. 1994a), methotrexate (ALEXANDER et al. 1994) or nocodazole (ALEXANDER et al. 1994) does not affect transduction implying that mitosis is not essential for AAV transduction. These findings are supported by demonstration of efficient AAV vector transduction of nonproliferating, respiratory epithelial cells (FLOTTE et al. 1994) and cellular populations that are normally quiescent including noncytokine stimulated hematopoietic progenitors (ZHOU SZ et al. 1994; Chatterjee et al., in press and submitted), primary human peripheral blood monocyte-macrophages (Chatterjee et al. 1994), and the brain (KAPLITT et al. 1994). Furthermore, AAV vectors can readily be designed for the expression of short, defined transcripts, a highly desirable feature when expressing either antisense transcripts, RNA decoys, or ribozymes (see below) (CHATTERJEE and WONG 1993). Importantly, for a gene therapeutic approach to AIDS, AAV vectors are unlikely to pseudotype with HIV to form chimeric viruses with either altered host range or pathogenicity (LUSSO et al. 1990). Finally, since its discovery in the mid-1960s, wild-type AAV has yet to be identified as a cause of disease in either animals or humans. On the contrary, there is evidence that infection with wild-type AAV inhibits transformation by bovine and human papillomaviruses, and the activated H-*ras* oncogene in vitro, while epidemiologic studies suggest that prior infection in humans may actually confer an oncoprotective effect (HERMONAT 1989, 1991, 1994; reviewed in ROMMELAERE and CORNELIS 1991; MAYOR 1993; SCHLEHOFER 1994). In addition, coinfection with wild-type AAV inhibits replication of several clinically important viruses, including adenovirus (CASTO et al. 1967; OSTROVE et al. 1981) and HIV (ANTONI et al. 1991; MENDELSON et al. 1992; RITTNER et al. 1992; OELZE et al. 1994), and cellular DNA amplification induced by HSV infection (BANTEL-SCHAAL and ZUR HAUSEN 1988). Inhibitory sequences have been mapped to the AAV p5 gene, which encodes Rep 78/68, potentially in combination with the inverted terminal repeats (ITRs). Such findings have important implications for the design of AAV-based antiviral vectors as they imply that specific AAV sequences and/or gene products may be directly inhibitory to the gene expression or replication of other viruses.

Recognized limitations of AAV vectors include a relatively small encapsidation capacity (about 5 kb total) and somewhat cumbersome vector excapsidation and purification systems. However, for the purposes discussed in this chapter, size constraints are not a major limitation as viral inhibitory transgenes are generally 1 kb or less in size so that AAV vectors could readily be designed to encode several. Furthermore, several laboratories are currently working toward the development of more reliable, efficient vector encapsidation and purification systems (YANG et al. 1994; FLOTTE et al. 1995; Chatterjee and Wong, unpublished). Efficient AAV vector transduction of primary cells including respiratory epithelial cells (rabbit, primates) (FLOTTE et al. 1993), brain (rat) (KAPLITT et al. 1994), and hematopoietic progenitors and their lineage committed progeny (mice, humans) (MURO-CACHO et al. 1992; CHATTERJEE et al. 1992b; ZHOU et al. 1994; GOODMAN et al. 1994; WALSH et al. 1994a,b; PODSAKOFF et al. 1994b; LU et al. 1994b; Chatterjee et al., submitted) have now been described by multiple laboratories.

4 Viral Inhibitory Molecules and Potential Targets

A variety of molecules have been examined for efficacy as antivirals and are most conveniently divided into RNA-based (including sense decoys, antisense transcripts, and catalytic ribozymes) and protein-based (negative transdominant peptides or proteins, single chain antibodies, inhibitors of viral trafficking, immunostimulatory cytokines, interferon or other antiviral species) molecules (Table 1). RNA-based approaches have the theoretical advantage of rapid onset of activity as they do not require translation and are nonimmunogenic but may be rapidly degraded by omnipresent nucleases. Protein-based inhibitors require transcription and translation for activity and may affect cellular processes or induce an immune response but are often potent antivirals. Relevant classes of

Table 1. Examples of gene transfer approaches to viral infection

	Target	Reference
Viral inhibitory molecules		
RNA-based		
Antisense	HIV/multiple sites	Rhodes et al. 1990
		Joshi et al. 1991
	HIV-1 TAR/PolyA	Chatterjee et al. 1992a
		Chuah et al. 1994
	HSV-1 ICP4	Wong et al. 1991
	MoMLV pac	Han et al. 1991
	MoMLV gag/pol	Sullenger et al. 1990a
Sense decoys	HIV TAR	Sullenger et al. 1990b
		Lisziewicz et al.1993
	HIV RRE	Lee et al. 1992
		Bevec et al. 1994
Ribozymes	HIV LTR	Weerasinghe et al. 1991
		Ojwang et al. 1992
		Yu et al. 1993
		Yamada et al. 1994
	HIV gag	Sarver et al. 1990
	HIV tat/rev	Zhou et al. 1994
	HPV-16 E6/E7 ORF	Lu et al. 1994
Protein-based		
Transdominant proteins	HIV *Tat*	Green et al. 1989
	HIV *Rev*	Malim et al. 1992
		Bahner et al. 1993
		Liu et al. 1994
	HIV *Gag*	Trono et al. 1989
		Smythe et al. 1994
	HIV *Env*	Buchschacher et al. 1992
	HTLV *Rex*	Bohnlein et al. 1991
	HSV VP16	Friedman et al. 1988
	HSV ICP4	Shepard et al. 1990
Intracellular antibodies	HIV gp120	Marasco et al. 1993
		Chen et al. 1994
	HIV *Rev*	Duan et al. 1994
Viral trafficking	HIV/sCD4	Morgan et al. 1990
	HIV/SCD4-ER retention	Buonocore and Rose 1990, 1993
	HIV/CD4-lysosome	Lin et al. 1993
Nucleases	MoMLV/Gag-staphylococcus nuclease	Natsoulis et al. 1995
Toxins	HIV/HSVtk	Brady et al. 1994
	HIV/diphtheria A chain toxin	Curiel et al. 1993
Other	Influenza/Mx1	Arnheiter et al. 1990
Immunomodulation	HIV/interferon	Bednarik et al. 1989
		Viellard et al. 1994
	Antigen presentation	Warner et al. 1991

viral inhibitory molecules, their proposed mechanisms of action, and examples of their use in the successful establishment of cellular viral resistance will be discussed briefly. Several, more extensive excellent reviews are available (James 1991; Wong and Chatterjee 1992; Yu et al. 1994; Shillitoe et al. 1994).

4.1 RNA-Based Inhibition of Gene Expression

4.1.1 Antisense RNA

Antisense transcripts perform important regulatory roles in prokaryotes (for reviews see GREEN et al. 1986; WAGNER and SIMONS 1994), have been developed as tools to specifically inhibit gene expression in eukaryotic cells (WEINTRAUB et al. 1985), and inhibit pathogenic viral replication in eukaryotic cells in vitro and in transgenic plants and animals in vivo (reviewed in VAN DER KROL et al. 1988; WONG and CHATTERJEE 1992). Although the exact mechanism of inhibition of gene expression is currently unknown, endogenously produced antisense RNA probably functions in a combinatorial fashion. Depending upon the target, sense:antisense hybrids are subject to rapid degradation by intracellular nucleases, disrupt inter- or intramolecular interactions, translation, intracellular processing or transport, and may induce actual mutations of the targeted transcript mediated by a cellular unwinding function (BASS and WEINTRAUB 1988) (for reviews see GREEN et al. 1986; VAN DER KROL et al. 1988; HELENE and TOULME 1990; EGUCHI et al. 1991). Intracellular expression of antisense transcripts complementary to essential viral transcripts has been effective in inhibiting Rous sarcoma virus (RSV) (CHANG and STOLTZFUS 1987), Moloney murine leukemia virus (MoMLV) (SULLENGER et al. 1990a; HAN et al. 1991), human adenovirus type 5 (Ad5) (MIROSHNICHENKO et al. 1989), HTLV-I (VON RUDEN and GILBOA 1989), and production HIV-1 (RHODES and JAMES 1990; SCZAKIEL and PAWLITA 1991; JOSHI et al. 1991; CHATTERJEE et al. 1992a). However, antisense RNA has not been universally effective in abrogating targeted gene expression, perhaps due to several factors including variability of antisense expression resulting from either low promoter strength or host chromosomal "position effects" flanking the inserted gene (ALLEN et al. 1988, SCHIMADA et al. 1991), inhibition of antisense:sense hybrid formation due to steric interference from RNA secondary structure, and extreme potency of the gene product such that residual quantities possess sufficient effector function for phenotypic gene expression (reviewed in VAN DER KROL et al. 1988).

4.1.2 Sense (Decoy) RNA

In contrast to the mechanisms proposed for antisense RNA-mediated inhibition of gene expression, intracellular expression of large quantities of truncated sense decoys containing *cis*-active regulatory regions compete with wild-type mRNA for essential, often rate-limiting, regulatory proteins, thereby inhibiting targeted gene expression and virus production. TAR, a requisite stem-loop sequence, has been identified at the 5′-end of essentially all HIV transcripts and is necessary for proper functioning of *Tat*, a virus-encoded protein that augments HIV gene expression (VAISHNAV and WONG-STAAL 1991). Thus intracellular synthesis of HIV-1 TAR containing transcripts interrupts requisite *Tat*-TAR interactions with subsequent inhibition of HIV long terminal repeat (LTR)-directed gene expression and virus production (SULLENGER et al. 1990b; JOSHI et al. 1991; LISZIEWICZ et al. 1993). A similar interaction of the HIV-1 *Rev* protein with *cis*-active elements contained within late HIV-1 encoded mRNAs (Rev responsive elements or RRE) is essential for the appropriate trafficking of unspliced or singly spliced transcripts (reviewed in VAISHNAV and WONG-STAAL 1991). Intracellular expression of RRE decoys disrupts *Rev*-RRE interaction and subsequent virus production (LEE et al. 1992; BEVEC et al. 1994). Efficient abrogation of targeted gene expression utilizing this strategy is dependent upon abundant expression of individual or concatemeric decoys.

4.1.3 Ribozymes

Ribozymes comprise a class of RNA molecules that were originally discovered through their unique autocatalytic activity (for reviews see CECH 1986; CASTANOTTO et al. 1994; POESCHLA and WONG-STAAL 1994). Hammerhead ribozymes, originally described from a class of plant pathogens, have subsequently been modified for the sequence-specific cleavage of RNA in *trans* (HASELOFF and GERLACH 1988). Theoretically any RNA transcript containing a specific cleavage sequence (classically in NUX, in which N can be A, C, G, or U, and X can be C, A, or U) can be cleaved with a hammerhead ribozyme targeted to the appropriate region by flanking complementary sequences. Recently, the constitutive intracellular expression of a hammerhead ribozyme targeting the HIV-1 *gag* transcript was both noncytotoxic and inhibited p24 Gag production by >99% following viral challenge (SARVER et al. 1990). Similarly, retroviral vectors encoding hammerhead ribozymes targeting 5′-leader or the first exon of *tat* or the exon common to both *tat* and *rev* conferred protection to CD4+ lymphocyte lines against HIV challenge (WEERASINGHE et al. 1991; ZHOU et al. 1994; reviewed in ZAIA et al. 1992).

In contrast to the hammerhead motif, hairpin ribozymes cleave immediately upstream of GUC sequences. A hairpin ribozyme targeting the HIV LTR and encoded within a retroviral vector conferred cellular resistance to laboratory and clinical isolates of HIV-1 in cell lines, in primary human CD4 T lymphocytes, and in umbilical cord blood derived hematopoietic progenitors and will soon be entering clinical trials (OJWANG et al. 1992; YU et al. 1993; YAMADA et al. 1994).

As ribozymes are targeted to the appropriate region of the transcript via flanking antisense sequences, they possess the advantages and disadvantages previously described for antisense RNA-mediated inhibition of gene expression. In addition, they have the advantage of destroying their target by cleavage and function in a catalytic rather than stoichiometric fashion. Thus, ribozyme expression may not have to be as robust as antisense expression for efficient inhibition of targeted gene expression. Finally, specific RNA binding proteins have recently been identified which enhance ribozyme-mediated target cleavage by enhancing the rate of product dissociation (BERTRAND and ROSSI 1994). Coexpression of these proteins with ribozymes may further enhance ribozyme-mediated inhibition of gene expression.

4.2 Protein-Based Inhibitors of Gene Expression

4.2.1 Transdominant Proteins

The development of transdominant inhibitory proteins (TIPs) is based upon the observation that essential regulatory proteins often consist of distinct modular domains which are necessary for both effector function and requisite intermolecular interactions (i.e., with another protein or nucleic acids) (reviewed in PTASHNE 1988). Thus, transdominant proteins consisting exclusively of one domain compete with the corresponding wild-type protein in the formation of functional multimeric complexes or for binding sites on DNA or RNA (HERSKOWITZ 1987). Utilizing this approach, mutant TIPs have been developed which functionally inhibit several HIV regulatory proteins including Tat (GREEN et al. 1989), Rev (MALIM et al. 1992), Gag (TRONO et al. 1989), and Env (BUCHSCHACHER et al. 1992), the HTLV-I Rex protein (BOHNLEIN et al. 1991), as well as HSV VP16 (FRIEDMAN et al. 1988) and ICP4 (SHEPARD et al. 1990). Importantly, intracellular expression of several of these transdominants has been associated with significant protection from cytopathogenicity and inhibition of viral production following challenge with their cognate virus (FRIEDMAN et al. 1988; TRONO et al. 1989; SHEPARD et al. 1990; MALIM et al. 1992; BUCHSCHACHER et al. 1992; BAHNER et al. 1993; LIEM et al. 1993; LIU et al. 1994; SMYTHE et al. 1994). Clinical trials of an HIV Rev transdominant are underway (NABEL et al. 1994; WOFFENDIN et al. 1994). However, TIPs are often based upon potent regulatory proteins and may retain regions which interact with and functionally disrupt important cellular factors. In addition, TIPs based upon viral regulatory proteins would be considered foreign and could induce an immune response. However, whether cellular expression of transdominant epitopes results in subsequent immune-mediated destruction of expressing cells has yet to be fully resolved.

4.2.2 Antibody Fragments

Intracellular expression of a specific antigen binding domain derived from a monoclonal antibody has been utilized to inhibit p21 ras protein expression in vitro (WERGE et al. 1990) and was suggested by FARAJI-SHADAN et al. (1990) as a potential gene therapeutic approach against HIV. Marasco developed a single-chain antibody, sFv105, derived from F105, an HIV-1 neutralizing antibody targeting the CD4+ binding site of HIV-1 gp120. Production of HIV-1 gp120 and infectious virus was blocked in COS cells (MARASCO et al. 1993) and in a human CD4+ lymphocyte line expressing sFv105 (CHEN et al. 1994). Duan developed a single chain antibody targeting HIV-1 Rev, a virus-encoded protein essential for cytoplasmic transport and translation of unspliced or singly spliced, late HIV mRNA. Cells expressing anti-Rev sFv were protected against challenge against both laboratory and clinical HIV-1 isolates (DUAN et al. 1994). Finally, human T lymphocyte lines (MOLT-3 and SupT-1) expressing a murine Fab with specificity to HIV reverse transcriptase restricted virus production by >80% following challenge with a variety of HIV strains (Maciejewski et al. 1995).

4.2.3 Nucleases

Recently Natsoulis and colleagues expressed a murine leukemia virus capsid (Gag)-staphylococcal nuclease fusion protein within chicken embryo fibroblasts (CEFs) (Natsoulis et al. 1995). This fusion protein was incorporated into virions following challenge with an amphotropic murine leukemia virus (Mo4070A), resulting in enzymatic cleavage of the viral genome, and a 30-to 100-fold reduction in virus production. Although intriguing, this strategy is dependent upon virion assembly and therefore requires some degree of virus replication and production to be effective. Such an approach would be problematic if virus-encoded products themselves were toxic to the cell.

4.2.4 Inhibitors of Viral Trafficking

Several gene transfer strategies to prevent viral infection involve methods which interfere with trafficking of virus particles or virus-encoded proteins. CD4 is the primary cellular receptor for HIV via its interaction with the viral envelope protein gp120 (Dalgleish et al. 1984). Morgan et al. (1990) demonstrated that cells transformed with a retroviral vector expressing soluble CD4 (sCD4) secreted sufficient quantities of sCD4 into the surrounding media to protect adjacent susceptible cells from subsequent challenge with a laboratory strain of HIV. However, the utility of this approach has been hampered by studies which demonstrate that very high concentrations of sCD4 were necessary to protect against actual clinical isolates of HIV-1 (Daar et al. 1990).

In contrast to *extra*cellular expression of sCD4, Buonocore and Rose (1990) utilized a vaccinia virus vector to express CD4 coupled to an endoplasmic reticulum (ER) retention signal (KDEL). Transiently-expressed HIV gp120 complexed with ER-retained CD4, thereby preventing transit to the cell surface and significantly inhibiting syncytia formation. Transduction of cells with a retroviral vector encoding this KDEL-CD4 construct conferred protection from subsequent HIV challenge with disappearance of infected cells 2 months postinfection (Buonocore and Rose 1993).

Similarly, intracellular expression of sCD4 coupled to signals for lysosomal trafficking resulted in diversion of sCD4-gp160 complexes to the lysosomal cellular compartment where they were proteolytically degraded. CEM T lymphoblastoid cell lines expressing the CD4-lysosome signal fusion proteins restricted HIV replication by 85%–95% by day 7 following virus challenge (Lin et al. 1993). The theoretical utility of these strategies stems from the utilization of the CD4 receptor as a common pathway for cellular entry of a serologically diverse group of pathogenic retroviruses, including both HIV-1 and HIV-2.

4.2.5 Toxins

Another strategy for a gene transfer approach to viral infections involves the incorporation of "suicide" genes encoding either enzymes which sensitize the cell to certain drugs or potent cellular toxins. One example of the former is the gene encoding herpes simplex thymidine kinase (HSV-*tk*) which sensitizes cells to the cytotoxic effects of ganciclovir, while an example of the latter is diphtheria toxin A chain (HARRISON et al. 1992; CURIEL et al. 1993). Cells are genetically engineered so that suicide genes are selectively and specifically induced following viral infection. For example, CD4 lymphocytes carrying a transgene encoding the diptheria toxin A chain under HIV LTR promoter control would express it following HIV infection resulting in the death of the cell. Similarly, CD4 lymphocytes carrying HSV-*tk* under HIV LTR control would be killed after HIV infection and exposure to ganciclovir (BRADY et al. 1994). Such an approach would limit opportunities of the targeted virus for replication, amplification, and subsequent spread within the host. However, one acknowledged disadvantage of this approach is the stringent control necessary for toxin expression. Many toxins are so potent that even very limited expression could result in cell death. Furthermore, limited infection with systemic release of diffusible molecules that *trans*-activate the promoter controlling toxin production could ultimately result in the death of all transgene containing cells. In addition, ganciclovir is the drug of choice for the treatment of cytomegalovirus infections, a common complication in AIDS patients, limiting the utility of the HSV-*tk* approach.

4.3 Other Antiviral Molecules/Immunomodulation

The studies of ARNHEITER et al. (1990), who developed influenza-resistant transgenic mice via pronuclear transfer of the Mx-1 resistance gene, and SALTER and CRITTENDEN (1989), who developed avian leukosis virus (ALV)-resistant transgenic chickens expressing the ALV envelope glycoprotein, are notable for employing naturally occurring mediators of viral resistance and demonstrating that they could be highly effective in vivo.

Intracellular expression of interferons have also been studied as an approach to antiviral gene transfer. Interferons are a class of potent, pleiotropic, broad-spectrum antiviral glycoproteins which are secreted by cells following a variety of stimuli, including viral infection and exposure to double-stranded RNA. Although the exact antiviral mechanisms are unknown, interferons interact with specific receptors at the cell surface and induce the synthesis of several enzymes which appear to be important for the subsequent development of an antiviral state. One of these enzymes, a protein kinase termed P1, phosphorylates the α subunit of eukaryotic protein synthesis initiation factor-2 (eIF-2α), thereby inhibiting protein synthesis, while another, (2′,5′)-oligoadenylate synthetase, induces an endonuclease that degrades single-stranded RNA (reviewed in STAEHELI 1990). CD4 positive human lymphocytes transformed with a retroviral vector containing the α-interferon gene under control of the inducible HIV LTR expressed elevated levels of α-interferon following HIV challenge and restricted virus production by greater than 95% relative to control cells (BEDNARIK et al. 1989). A similar approach was utilized by SEIF et al. (1991), who transfected murine cells with a plasmid conferring G418 resistance and containing the murine β-interferon gene under control of the murine H-2K^b major histocompatibility (MHC) promoter. Clonally derived cells expressing β-interferon displayed enhanced resistance to vesicular stomatitis, encephalomyocarditis, and Semliki Forest viruses in comparison to control cells. Similarly, target cells transduced with a retroviral vector encoding human β-interferon were protected against either retroviral vector transduction or HIV-1 infection (VIEILLARD et al. 1994). Augmented interferon expression offers the advantages of a broad antiviral spectrum associated with the ability to confer protection to untransduced neighboring cells. However, the pleiotropic effects of interferons, including their broad effects upon the immune system and cellular proliferation, associated with the recognized in vivo toxicity of continuous high concentrations of interferon (reviewed in GRESSER 1986), would necessitate tightly regulated expression and might hamper this interesting application in vivo.

Gene transfer approaches have also been developed which stimulate humoral and/or cellular antiviral immune responses. Murine fibroblasts transduced with a retroviral vector encoding the HIV-III$_B$ envelope (Env) protein induced both MHC-I restricted, CD8+ cytotoxic responses directed against Env-expressing targets as well as Env-specific antibody responses (WARNER et al. 1991). This approach has also shown promise in generating antiviral immune responses in nonhuman primates and is currently undergoing clinical evaluation (GALPIN et al. 1994).

Finally, ROBERTS recently described retroviral vector mediated transfer of a universal (MHC-unrestricted) chimeric T cell receptor gene targeting HIV antigens into CD8+ T cells (ROBERTS et al. 1994). This strategy was employed to generate a rapid cytotoxic response directed against HIV-infected cells by circumventing MHC-restricted antigen presentation. The cytoplasmic domain of the invariant chain of the T cell receptor (TCR) was coupled to either the extracellular domain of the human CD4 receptor (CD4-UR), the primary receptor for HIV, or a single chain antibody against the gp41 region of the HIV Env (SAb-UR). Thus, engagement of either CD4-UR or SAb-UR with its cognate antigen should result in immediate lymphocyte activation, without the need for antigen presentation or costimulatory signals. Primary human CD8+ lymphocytes transduced with retroviral vectors encoding either CD4-UR or SAb-UR displayed both specific proliferative responses after exposure to HIV Env and enhanced cytolysis of HIV infected cells.

5 Examples of Gene Transfer Approaches to Viral Infections Using Adeno-associated Virus-Based Vectors

5.1 Herpes Simplex Virus

Although generally effective in acute infections, current drug treatment is not effective against latent HSV and thus is not curative. Furthermore, drug resistant HSV strains are being reported with increasing frequency, particularly in immunosuppressed individuals such as AIDS patients (BALFOUR et al. 1994). Thus, novel approaches to the control of HSV infection are continually being sought.

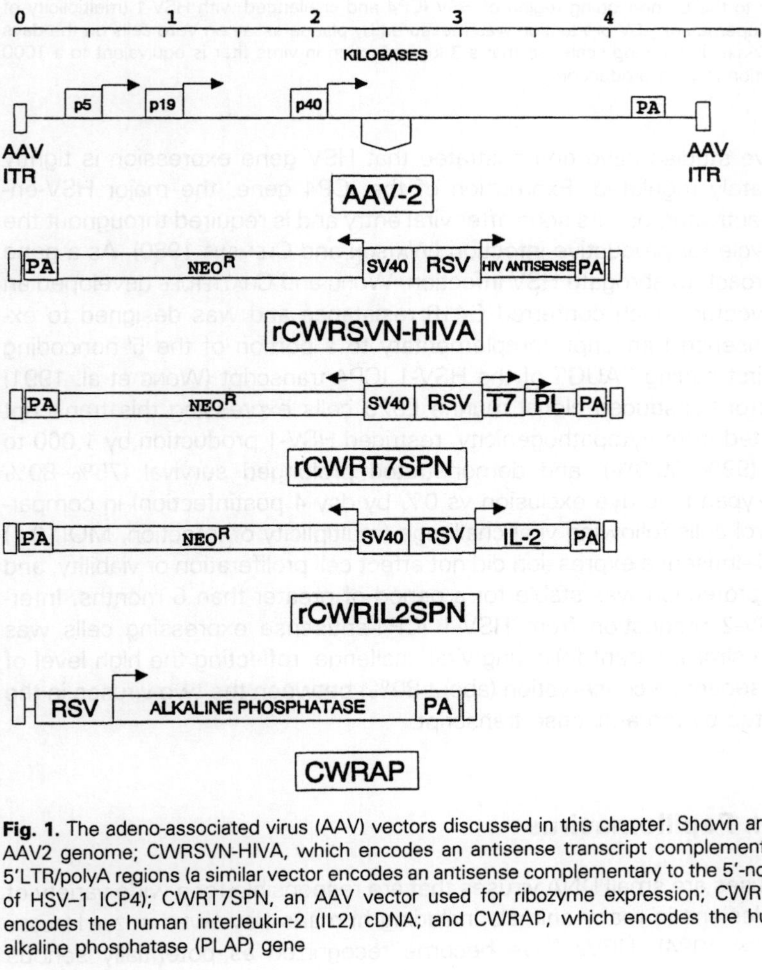

Fig. 1. The adeno-associated virus (AAV) vectors discussed in this chapter. Shown are the wild-type AAV2 genome; CWRSVN-HIVA, which encodes an antisense transcript complementary to the HIV 5'LTR/polyA regions (a similar vector encodes an antisense complementary to the 5'-noncoding region of HSV-1 ICP4); CWRT7SPN, an AAV vector used for ribozyme expression; CWRIL2SPN, which encodes the human interleukin-2 (IL2) cDNA; and CWRAP, which encodes the human placental alkaline phosphatase (PLAP) gene

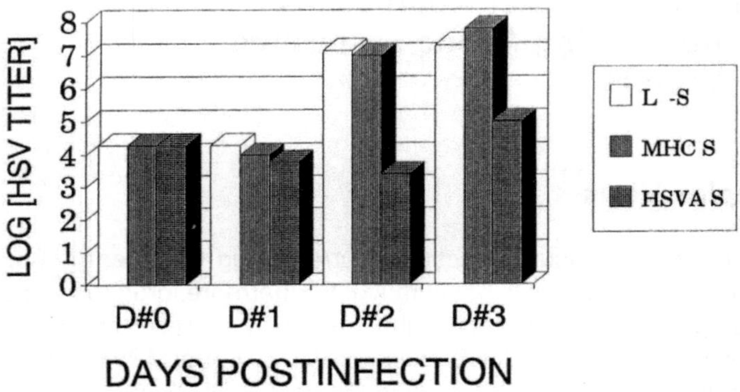

Fig. 2. Restriction of herpes simplex virus (HSV)-1 production in clonally derived murine L929 cells transduced with an adeno-associated virus (AAV) vector encoding an antisense transcript complementary to the 5'- noncoding region of HSV ICP4 and challenged with HSV-1 (multiplicity of infection 0.1). Supernatant HSV production was measured by plaque assay on Vero cells on the days indicated. The *Y-axis* is on a log scale, so that a 3 log reduction in virus titer is equivalent to a 1000 (10^3)-fold reduction in virus production

Extensive studies have demonstrated that HSV gene expression is tightly and coordinately regulated. Expression of the ICP4 gene, the major HSV-encoded *trans*-activator, occurs soon after viral entry and is required throughout the replicative cycle for productive infection (WATSON and CLEMENS 1980). As a gene transfer approach to abrogate HSV infection, WONG and CHATTERJEE developed an AAV-based vector which conferred G418 resistance and was designed to express an antisense transcript complementary to a portion of the 5'-noncoding leader and first coding "AUG" of the HSV-1 ICP4 transcript (WONG et al. 1991) (Fig. 1). Vector-transduced clonal murine L929 cells expressing this transcript were protected from cytopathogenicity, restriced HSV-1 production by 1,000 to 10,000-fold (99%–99.9%), and demonstrated prolonged survival (75%–80% viability by trypan blue dye exclusion vs 0% by day 4 postinfection) in comparison to control cells following viral challenge (multiplicity of infection, MOI, 0.1) (Fig. 2). ICP4-antisense expression did not affect cell proliferation or viability, and the level of protection was stable for a period of greater than 6 months. Interestingly, HSV-2 production from HSV-1 ICP4-antisense expressing cells was inhibited to a similar extent following viral challenge, reflecting the high level of nucleic acid sequence conservation (about 80%) between the two viruses in the region targeted by the antisense transcript.

5.2 Human Papillomavirus

Papillomaviruses are small DNA viruses that are responsible for a wide variety of proliferative lesions in many animals including humans (reviewed in ZUR HAUSEN and DE VILLIERS 1994). HPVs have become recognized as potentially serious pathogens due to their high prevalence, increasing incidence, and identification as important cofactors in the development of cancer, particularly anogenital carcinomas. HPV-16 and HPV-18 DNA have been found in over 85% of invasive squamous cell carcinomas of the uterine cervix. Importantly, protective vaccines are unavailable and current therapies, including topical podophyllin, and systemic or intralesional interferon, are of uncertain value in the prevention of malignant transformation.

Elegant mutational studies utilizing cloned subgenomic DNA fragments have defined products from the HPV E6 and E7 open reading frames (ORFs) as essential for viral transformation (reviewed in HUIBREGTSE and SCHEFFNER 1994). DNA encoding HPV-16 and 18 E6 and E7 ORFs transform epithelial cells in vitro, including human cervical epithelial cells. Mechanistically, the E6 product from transforming HPV types has been shown to physically associate with the cellular tumor suppressor gene product p53 (WERNESS et al. 1990), resulting in its rapid catabolism, possibly via the protein catabolic ubiquitin pathway. Likewise, the E7 ORF product binds to the cellular retinoblastoma tumor suppressor gene product (105RB), disrupting its function (DYSON et al. 1989). Thus, disruption of endogenous cellular antitransformation protective mechanisms by virus-encoded genes may play a major role in HPV-mediated transformation.

LU et al. designed hammerhead ribozymes which cleave transcripts encoding HPV-16 E6 and E7 ORFs in proximity to the translational start sites AUG of their respective proteins. These regions are known to be effective targets for antisense transcript-mediated inhibition of gene expression (LU et al. 1994a). Ribozyme target sites were specifically chosen within exons common to all known E6 and E7 encoding transcripts, with preference given to sites with high cleavage efficiencies according to the hierarchy determined by PERRIMAN et al. (1992). Thus, cleavage was designed to occur immediately 3′ to UUC and GUA sequences corresponding to nucleotides 110 and 558 of the viral genomic DNA, respectively. Ribozyme-mediated transcript cleavage at 110 would remove the translational start codon (AUG) from E6 transcripts as well as the 5′7-methyl guanosine cap necessary for efficient translation from E6 and E7 encoding transcripts. Likewise, transcript cleavage at 558 would remove the 5′7-methyl guanosine cap from E7 transcripts and the translational terminator (UAA) and polyadenylation signal from E6 transcripts. To further augment inhibition of targeted gene expression, the flanking antisense sequences, which confer ribozyme sequence specificity, are complementary to the AUG translational start codons for E6 and E7 transcripts, disrupting E6 and E7 translation. Oligonucleotides corresponding to these ribozymes were synthesized and inserted into CWRT7:SVN, an AAV-derived vector. Ribozyme transcription from this vector is under control of both the highly active RSV LTR and bacteriophage T7 promoters (Fig. 1). E6 and E7 ribozymes were transcribed in vitro, and cleavage of cognate targets occurred efficiently under physiologic conditions. Studies of the potential antiproliferative effects of transduction of HPV-transformed cells with E6/E7 ribozyme encoding AAV vectors are currently underway.

5.3 Human Immunodeficiency Virus

The AIDS epidemic has had a major social, medical, ethical, and economic impact worldwide, with projections of incidence and prevalence rates of potentially catastrophic proportions by the turn of the century (reviewed in SCHNITTMAN and FAUCI 1994). AIDS is a complex, ultimately fatal multisystem disease caused by infection with HIV and characterized by widespread clinical ramifications resulting from progressive deterioration of the immune system. Despite extensive investigation, chemotherapeutic control of HIV infection remains elusive. Currently available drugs such as zidovudine (AZT), didanosine (ddl), or zalcitabine (ddC) are associated with significant toxicity and are not curative, while drug-resistant strains are rapidly evolving (RICHMAN 1993). Furthermore, progress in AIDS vaccine development has been slow due to mutatability of the HIV genome. In contrast, specific strategies for the control of HIV-1 infection designed to interrupt requisite viral functions at the molecular level, as described above, appear promising in vitro. Ultimately, the goal of such approaches is the stable genetic alteration of the normal cellular targets of HIV infection rendering them resistant to productive infection. In this strategy, HIV infection itself exerts a selective pressure for outgrowth of resistant cells, ultimately preventing the catastrophic immunologic collapse that occurs in end stage AIDS. The potential of this approach is supported by an animal model of murine AIDS in which transplantation with virus-resistant marrow cells was somewhat protective against development of disease (NAKAGAWA et al. 1994).

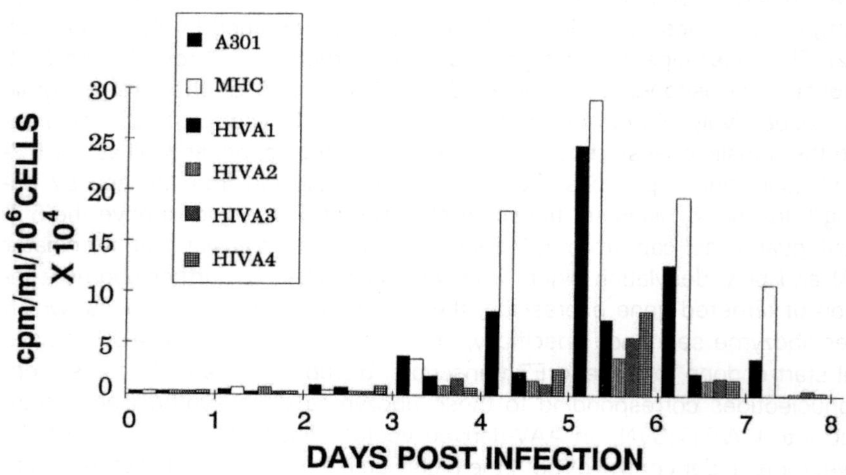

Fig. 3. Inhibition of reverse transcriptase (RT) activity from nonclonal, G418-resistant, CD4+ T lymphocytes (A3.01) transduced with an adeno-associated virus (AAV) vector encoding an antisense complementary to the HIV LTR/polyA region and challenged with HIV-1 (multiplicity of infection 1). Shown are the results of RT assays following HIV challenge of unmanipulated control cells (A3.01), control AAV vector-transduced (MHC), and four individually transduced and bulk G418-selected cells lines (HIVA1–A4). (Modified from CHATTERJEE et al. 1992a)

CHATTERJEE et al. (1992a) constructed an AAV-based vector encoding an antisense transcript complementary to both the HIV-1 TAR region and polyadenylation signal, common to all HIV strains, under control of the RSV LTR (Figure 1). This vector, termed CWRSVN-HIVA, also encoded neomycin phosphotransferase and thus conferred cellular resistance to G418. Transduced, clonally derived, G418-resistant 293 cells expressing this antisense transcript specifically restricted chloramphenicol acetyltransferase (CAT) expression directed from the HIV-1 LTR as well as HIV replication following transfection with HIV_{IIIB}, an infectious molecular clone. Importantly, similarly transduced, nonclonal but G418-resistant, human CD4 T cell lines (H9 and A3.01) expressing the antisense RNA were protected with a 1000-fold reduction of titratable virus production after challenge with HIV_{IIIB} even at a MOI of 1 (Fig. 3). These cells were also protected, although to a lesser extent, from challenge with simian immunodeficiency virus (SMMH4i), a serologically and genetically diverse lentivirus which shares some nucleotide homology with HIV-1 at the TAR and polyadenylation regions. However, similarly derived CD4 cells transduced with control AAV vectors encoding irrelevant transcripts, including the HSV-1 ICP4 antisense, were not protected from HIV challenge implying that the inhibitory effect was not due to nonspecific effects. No significant contamination of vector stocks with wild-type AAV was noted. Cellular protection has also been demonstrated following transduction with retroviral vectors encoding similar TAR/polyA antisense transcripts (CHUAH et al. 1994). AAV vector-mediated delivery of other HIV inhibitory molecules, including hammerhead and hairpin ribozymes, TAR and RRE decoys, single chain antibodies, and Rev transdominant proteins, is actively being investigated in our laboratory and elsewhere.

For gene therapy of AIDS, either mature, differentiated targets of HIV infection (CD4+ T lymphocytes, monocyte-macrophages) or their progenitors (hematopoietic stem cells), could be genetically modified to render them resistent to subsequent HIV challenge. To determine the efficiency of AAV-mediated gene transfer, we developed a series of vectors which encode thermostable placental alkaline phosphate (PLAP) and which allow rapid determination of transduction efficiencies by in situ histochemical staining (Fig. 1). Up to 65% of primary human $CD4^+$ T lymphocytes, and 70%–95% of primary peripheral blood monocyte-macrophages demonstrated specific PLAP expression following transduction with these AAV vectors. Sustained transgene expression was observed up to 8 weeks posttransduction for T lymphocytes and up to 5 weeks for macrophages, the length of the study (CHATTERJEE et al. 1994, in press). Primary human marrow-derived hematopoietic cells enriched for CD34 antigen expression, a marker for more primitive progenitors, demonstrated transduction efficiencies of >70%, detectable PLAP expression for 5 weeks, and evidence of vector sequences for >6 weeks after culture initiation (LU et al. 1994b; CHATTERJEE et al. 1992b, in press, submitted). These studies support the feasibility of AAV vectors for gene transfer to these critical hematopoietic populations and a pivotal role for the use of these vectors for the gene therapy of AIDS.

As previously mentioned, the AAV p5 product (Rep 78/68) inhibits both gene expression arising from the HIV-LTR as well as HIV replication (ANTONI et al. 1991; RITTNER et al. 1992; MENDELSON et al 1992; OELZE et al. 1994). Inhibition of HIV-LTR gene expression occurs at the transcriptional level, possibly resulting from the stimulation of an inhibitory region with the LTR itself (OELZE et al. 1994). Although the inhibitory effects of AAV2 infection upon HIV replication were modest (20% reduction in reverse transcriptase activity following AAV2 and Ad2 coinfection) (MENDELSON et al. 1992), it might be possible to construct vectors which express the HIV inhibitory function(s) of Rep 78/68 in conjuction with other viral inhibitory molecules to potentiate the effect.

5.4 Immunomodulation

Adeno-associated virus vectors could be used to potentiate or induce immune responses by several pathways. AAV vectors encoding interferons, cytokines, or lymphokines could be used to stimulate immunity nonspecifically. For example, an AAV vector encoding the human interleukin 2 (IL-2) cDNA under RSV LTR promoter control is currently undergoing evaluation in our laboratory (Fig. 1). IL-2 is a pleiotropic cytokine which promotes the proliferation of antigen stimulated T cells, natural killer (NK) and lymphokine activated killer (LAK) cells and thus has a potent effect upon the immune system (reviewed in BRUTON and KOELLER 1994). Cells transduced with this vector produce significant quantities of IL-2 by both antigenic (ELISA) (Fig. 4) and functional assays (cytotoxic T lymphocyte proliferation bioassay). Expression of IL-2, interferon, or other broadly active immunostimulatory cytokine could elicit a significant antiviral response.

Conversely, AAV vectors encoding specific protein/peptide epitopes could be used to induce specific immune responses directed against a distinct viral target. For example, murine fibroblasts have been transduced in vitro with a

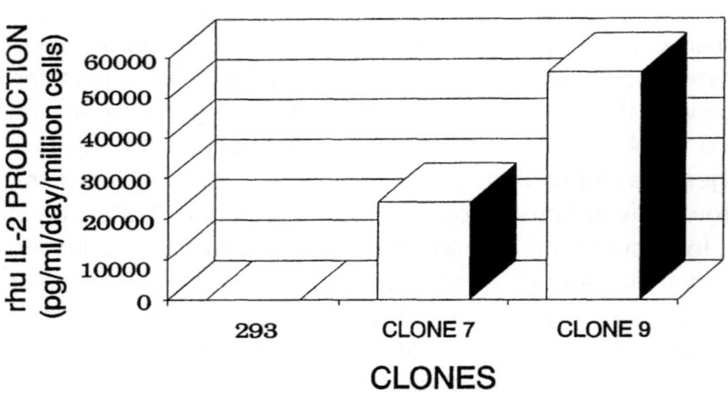

Fig. 4. Expression of rHuIL2 from clonally derived 293 cells transduced with an adeno-associated virus (AAV)

retroviral vector encoding the HIV-1 *env* gene. Although subsequently irradiated to prevent proliferation, these cells were still capable of antigen presentation and production of both cytotoxic and humoral anti-HIV immune responses in vivo (WARNER et al. 1991). Because of their ability to efficiently transfer genes into nonproliferating cells, AAV vectors may be more effective in genetically modifying professional antigen presenting cells, thereby generating a more potent antiviral immune response.

6 Perspectives, Problems and Future Considerations

Adeno-associated virus-based vectors appear highly promising as vehicles for efficient gene transfer of a variety of antiviral molecules. What questions remain? The mechanisms and overall efficiency of AAV vector integration into populations of primary cells are still undefined; both integrated and episomal vector forms have been identified (FLOTTE et al. 1994; Chatterjee et al., in press). Furthermore, essentially all studies to date involving AAV-mediated gene transfer have been done in tissue culture; in vivo experience is relatively limited. Flotte employed an AAV vector to introduce the cystic fibrosis transmembrane regulator (CFTR) into rabbit bronchial cells in vivo as a model for the gene therapy of cystic fibrosis. CFTR expression could be detected for 6 months after transduction implying stable genetic modification of respiratory epithelium (FLOTTE et al. 1993). Kaplitt injected AAV vectors encoding β-galactosidase and tyrosine hydroxylase into specific areas of the postmitotic rat brain and documented transgene expression for 3 and 4 months postinjection, respectively (KAPLITT et al. 1994). PODSAKOFF used an AAV vector encoding PLAP to transduce murine hematopoietic progenitors ex vivo, prior to transplantation into syngeneic lethally irradiated recipients. PLAP expression could be detected for 6 months posttransplantation, while vector sequences could be detected by PCR in peripheral blood, spleen, marrow, and thymus for 6–7.5 months after reconstitution (PODSAKOFF et al. 1994b). Vector specific sequences were identified at 11 weeks posttransplantation in three of four CB17 scid/scid mice transplanted with human umbilical cord blood cells transduced with an AAV vector encoding the gene for Fanconi anemia complementation group C (FACC) (WALSH et al. 1994b). Although one group has reported difficulty in transducing primary cells in vitro (HALBERT et al. 1995), results from several independent laboratories using different AAV vectors, transgenes, and animal models support the potential role of AAV vectors for gene transfer in vivo.

Will the high transduction efficiencies seen in vitro translate to better gene transfer in vivo? Does prior immunity affect vector transduction or the survival of AAV vector-transduced cells? Do *rep78/68* deleted AAV vectors integrate randomly, and if not, where do they integrate? Will expression from AAV-encoded transgenes be stable, or will it decline over time? Can a better encapsidation/

vector production system be developed? We and others are actively involved in trying to answer these questions using both in vitro and in vivo studies. The NIH Recombinant DNA Advisory Committee approved the first human clinical protocol using an AAV vector in September 1994. In this study, an AAV vector encoding the CFTR cDNA will be used to determine transduction efficiencies, stability of gene expression, and safety issues following vector instillation into subsegmental regions of the lung in patients with mild cystic fibrosis. Our laboratory is actively studying AAV-mediated gene transfer to hematopoietic cells in animal models with the goal of advancing to human clinical trials. Based on our results, we believe that AAV vectors can play a pivotal role in gene transfer approaches to the treatment of HIV and other serious viral infections.

Acknowledgements. We thank members of our laboratory for their many helpful comments and suggestions. We also thank John A. Zaia, Stephen J. Forman, John Kovach and the members of the City of Hope Bone Marrow transplantation team for their help and continued support. This work was supported in part by grants CA59308, CA33572, AI25959 from the National Institutes of Health.

References

Alexander IE, Russell DW, Miller AD (1994) DNA-damaging agents greatly increase the transduction of nondividing cells by adeno-associated virus vectors. J Virol 68: 8282–8287

Allen ND, Cran DG, Barton SC, Hettle S, Reik R, Surani MAH (1988) Transgenes as probes for active chromosomal domains in mouse development. Nature 333: 852–855

Anderson WF (1992) Human Gene Therapy. Science 256: 808–813

Antoni BA, Rabson AB, Miller IL, Trempe JP, Chejanovsky N, Carter BJ (1991) Adeno-associated virus *rep* protein inhibits human immuno-deficiency virus type 1 production in human cells. J Virol 65: 396–404

Arnheiter H, Skuntz S, Noteborn M, Chang S, Meler E (1990) Transgenic mice with intracellular immunity to influenza virus. Cell 62: 51–61

Bahner I, Zhou C, Yu XJ, Hao QL, Guatelli JC, Kohn DB (1993) Comparison of trans-dominant inhibitory mutant human immunodeficiency virus type 1 genes expressed by retroviral vectors in human T lymphocytes. J Virol 67: 3199–3207

Balfour HH Jr, Benson C, Braun J, Cassens B, Erice A, Friedman-Kien A, Klein T, Polsky B, Safrin S (1994) Management of acyclovir-resistant herpes simplex and varicella-zoster virus infections. J Acquir Immune Defic Syndr 7: 254–260

Baltimore D (1988) Intracellular immunization. Nature 335: 395–396

Bantel-Schaall U, zur Hausen H (1988) Adeno-associated viruses inhibit SV40 DNA amplification and replication of herpes simplex virus in SV40-transformed hamster cells. Virology 164: 64–74

Bass BL, Weintraub H (1988) An unwinding activity that covalently modifies its double-stranded RNA substrate. Cell 55: 1089–1098

Bednarik DP, Mosca JD, Raj NBK, Pitha PM (1989) Inhibition of human immunodeficiency virus (HIV) replication by HIV-trans-activated alpha-2 interferon. Proc Natl Acad Sci USA 86: 4958–4962

Berns KI, Bohenzky RA (1987) Adeno-associated viruses: an update. Adv Virus Res 32: 243–306

Berns KI, Pinkerton TC, Thomas GF, Hoggan MD (1975) Detection of adeno-associated virus (AAV)-specific nucleotide sequences in DNA isolated from latently infected Detroit 6 cells. Virology 68: 556–560

Bevec D, Volc-Platzer B, Zimmermann K, Dobrovnik M, Hauber J, Veres G, Bohnlein E (1994) Constitutive expression of chimeric neo-Rev response element transcripts suppresses HIV-1 replication in human CD4+ T lymphocytes. Hum Gene Ther 5: 193–201

Bertrand EL, Rossi JJ (1994) Facilitation of hammerhead ribozyme catalysis by the nucleocapsid protein of HIV-1 and the heterogeneous nuclear ribonucleoprotein A1. EMBO J 13: 2904–2912

Blaese RM (1993) Development of gene therapy for immunodeficiency: adenosine deaminase deficiency. Pediatr Res 33 [Suppl]: S49–S53
Bohnlein S, Pirker FP, Hofer L, Zimmermann K, Bachmayer H, Bohnlein E, Hauber J (1991) Transdominant repressors for human T-cell leukemia virus type I *rex* and human immunodeficiency virus type 1 *rev* function. J Virol 65: 81–88
Brady HJ, Miles CG, Pennington DJ, Dzierzak EA (1994) Specific ablation of human immunodeficiency virus Tat-expressing cells by conditionally toxic retroviruses. Proc Natl Acad Sci USA 91: 365–369
Brenner MK, Rill DR, Holladay MS, Heslop HE, Moen RC, Buschle M, Krance RA, Santana VM, Anderson WF, Ihle JN (1993) Gene marking to determine whether autologous marrow infusion restores long-term haemopoiesis in cancer patients. Lancet 342: 1134–1137
Bruton JK, Koeller JM (1994) Recombinant interleukin-2. Pharmacotherapy 14: 635–656
Buchschacher GL, Freed EO, Panganiban AT (1992) Cells induced to express a human immunodeficiency virus type 1 envelope gene mutant inhibit the spread of wild-type virus. Hum Gene Ther 3: 391–397
Buonocore L, Rose JK (1990) Prevention of HIV-1 glycoprotein transport by soluble CD4 retained in the endoplasmic reticulum. Nature 345: 625–628
Buonocore L, Rose JK (1993) Blockade of human immunodeficiency virus type 1 production in CD4+ T cells by an intracellular CD4 expressed under control of the viral long terminal repeat. Proc Natl Acad Sci USA 90: 2695–2699
Carter RB, Abrams-Ogg AC, Dick JE, Kruth SA, Valli VE, Kamel-Reid S, Dube ID (1992) Autologous transplantation of canine long-term marrow culture cells genetically marked by retroviral vectors. Blood 79: 356–364
Castanotto D, Rossi JJ, Sarver N (1994) Antisense catalytic RNAs as therapeutic agents. Adv Pharm 25: 289–317
Casto BC, Armstrong JA, Atchinson RW, Hammon WMcD (1967) Studies on the relationship between adeno-associated virus type 1 (AAV-1) and adenoviruses. II. Inhibition of adenovirus plaques by AAV; its nature and specificity. Virology 33: 452–458
Cech TR, Bass BL (1986) Biological catalysis by RNA. Annu Rev Biochem 55: 599–629
Chang L-J, Stoltzfus CM (1987) Inhibition of Rous sarcoma virus replication antisense RNA. J Virol 61: 921–924
Chatterjee S, Wong KK Jr (1993) Adeno-associated viral vectors for the delivery of antisense RNA. Methods (a companion to methods. Enzymol) 5: 51–59
Chatterjee S, Johnson PR, Wong KK Jr (1992a) Dual target inhibition of HIV-1 in vitro by means of an adeno-associated virus antisense vector. Science 258: 1485–1488
Chatterjee S, Wong KK, Podsakoff G, Zaia J, Forman S (1992b) Adeno-associated virus vectors for high efficiency gene transfer into primary hematopoietic cells. Blood 80: 167a
Chatterjee S, Podsakoff G, Wong KK Jr (1994) Gene transfer into terminally differentiated primary human peripheral blood-derived mononuclear cells by adeno-associated virus. Blood 84: 360a
Chatterjee S, Wong KK Jr, Lu D, Permana P, Podsakoff G. Novel approaches for efficient gene transfer into hematopoietic progenitor cells: the use of adeno-associated virus vectors. Bone Marrow Transplant (in press)
Chatterjee S, Podsakoff G, Forman SJ, Wong KK Jr (1995) Adeno-associated virus-mediated gene transfer into CD34+ primary human hematopoietic progenitor cells is highly efficient and stable (submitted)
Chen SY, Khouri Y, Bagley J, Marasco WA (1994) Combined intra- and extracellular immunization against human immunodeficiency virus type 1 infection with a human anti-gp120 antibody. Proc Natl Acad Sci USA 91: 5932–5936
Chuah MKL, Vandendriessche T, Chang H-K, Ensoli B, Morgan RA (1994) Inhibition of human immunodeficiency virus type-1 by retroviral vectors expressing antisense-TAR. Hum Gene Ther 5: 1467–1475
Curiel TJ, Cook DR, Wang Y, Hahn BH, Ghosh SK, Harrison GS (1993) Long-term inhibition of clinical and laboratory human immunodeficiency virus strains in human T-cell lines containing an HIV-regulated diphtheria toxin A chain gene. Hum Gene Ther 4: 741–747
Cuypers HT, Selten G, Quint W, Zijlstra M, Maandag ER, Boelens W, van Wezenbeek P, Melief C, Berns A (1984) Murine leukemia virus-induced T-cell lymphomagensis: integration of proviruses in a distinct chromosomal region. Cell 37: 141–150
Daar ES, Li XL, Moudgil T, Ho DD (1990) High concentrations of recombinant soluble CD4 are required to neutralize primary human immunodeficiency virus type 1 isolates. Proc Natl Acad Sci USA 87: 6574–6578

Dalgleish AG, Beverley PCL, Clapham PR, Crawford DH, Greaves MF, Weiss RA (1984) The CD4 (T4) antigen is an essential component of the receptor for the AIDS retrovirus. Nature 312: 763–767
Donahue RE, Kessler SW, Bodine D, McDonagh K, Dunbar C, Goodman S, Agricola B, Byrne E, Raffeld M, Meon R, Backer J, Zsebo KM, Nienhuis AW (1992) Helper virus induced T cell lymphoma in nonhuman primates after retroviral mediated gene transfer. J Exp Med 176: 1125–1135
Duan L, Bagasra O, Laughlin MA, Oakes JW, Pomerantz RJ (1994) Potent inhibition of human immunodeficiency virus type 1 replication by an intracellular anti-rev single chain antibody. Proc Natl Acad Sci USA 91: 5075–5077
Dunbar CE, Cottler-Fox M, O'Shaughnessy JA, Doren S, Carter C, Berenson R, Brown S, Moen RC, Greenblatt J, Stewart FM, Leitman SF, Wilson WH, Cowan K, Young NS, Nienhuis AW (1995) Retrovirally marked CD34-enriched peripheral blood and bone marrow cells contribute to long-term engraftment after autologous transplantation. Blood 85: 3048–3057
Dyson N, Howley PM, Münger K, Harlow E (1989) The human papilloma virus-16 oncoprotein is able to bind to the retinoblastoma gene product. Science 243: 934–937
Eguchi Y, Itoh T, Tomizawa J (1991) Antisense RNA. Annu Rev Biochem 60: 631–652
Faraji-Shadan F, Stubbs JD, Bowman PD (1990) A putative approach for gene therapy against human immunodeficiency virus (HIV). Med Hypotheses 32: 81–84
Field AK, Biron KK (1994) The end of innocence revisited: resistance of herpesviruses to antiviral drugs. Clin Microbiol Rev 7: 1–13
Flotte TR, Afione SA, Conrad C, McGrath SA, Solow R, Oka H, Zeitlin PL, Guggino WB, Carter BJ (1993) Stable in vivo expression of the cystic fibrosis transmembrane conductance regulator with an adeno-associated virus vector. Proc Natl Acad Sci USA 90: 10613–10617
Flotte TR, Afione SA, Zeitlin PL (1994) Adeno-associated virus vector gene expression occurs in nondividing cells in the absence of vector DNA integration. Am J Respir Cell Mol Biol 11: 517–521
Flotte TR, Barraza-Ortiz X, Solow R, Afione SA, Carter BJ, Guggino WB (1995) An improved system for packaging recombinant adeno-associated virus vector capable of in vivo transduction. Gene Therapy 2: 29–37
Friedman AD, Triezenberg SJ, McKnight SL (1988) Expression of a truncated viral transactivator selectively impedes lytic infection by its cognate virus. Nature 335: 452–454
Galpin JE, Casciato DA, Richards SB (1994) A phase I clinical trial to evaluate the safety and biological activity of HIV-IT (TAF) (HIV-1IIIBenv-transduced, autologous fibroblasts) in asymptomatic HIV-1 infected subjects. Hum Gene Ther 5: 997–1017
Goodman S, Xiao X, Donahue RE, Moulton A, Miller J, Walsh C, Young NS, Samulski RJ, Nienhuis AW (1994) Recombinant adeno-associated virus-mediated gene transfer into hematopoietic progenitor cells. Blood 84: 1492–1500
Green M, Ishino M, Loewenstein PM (1989) Mutational analysis of HIV-1 *Tat* minimal domain peptides: identification of trans-dominant mutants that suppress HIV-LTR-driven gene expression. Cell 58: 215–223
Green PJ, Pines O, Inouye M (1986) The role of antisense RNA in gene regulation. Annu Rev Biochem 55: 569–597
Gresser I (1986) Interferon-induced diseases. In: Notkins AL, Oldstone MBA (eds) Concepts in viral pathogenesis II. Springer, Berlin Heidelberg New York, pp 232–242
Halbert CL, Alexander IE, Wolgamot GM, Miller AD (1995) Adeno-associated virus vectors transduce primary cells much less efficiently than immortalized cells. J Virol 69: 1473–1479
Han L, Yun JS, Wagner TE (1991) Inhibition of Moloney murine leukemia virus-induced leukemia in transgenic mice expressing antisense RNA complementary to the retroviral packaging sequences. Proc Natl Acad Sci USA 88: 4313–4317
Harrison GS, Long CJ, Curiel TJ, Maxwell F, Maxwell IH (1992) Inhibition of human immunodeficiency virus-1 production resulting from transduction with a retrovirus containing an HIV-regulated diphtheria toxin A chain gene. Hum Gene Ther. 3: 461–469
Hartung S, Jaenisch R, Breindl M (1986) Retrovirus insertion inactivates mouse al(I) collagen gene by blocking initiation of transcription. Nature 320: 365–367
Haseloff J, Gerlach WL (1988) Simple RNA enzymes with new and highly specific endoribonuclease activities. Nature 334: 585–591
Hayward WS, Neel BG, Astrin SM (1981) Activation of a cellular oncogene by promoter insertion in ALV-induced lymphoid leukosis. Nature 290: 475–480
Helene C, Toulme J-J (1990) Specific regulation of gene expression by antisense, sense and antigene nucleic acids. Biochim Biophys Acta 1049: 99–125

Hermonat PL (1989) The adeno-associated virus *rep* 78 gene inhibits cellular transformation induced by bovine papillomavirus. Virology 172: 253–261

Hermonat PL (1991) Inhibition of H-ras expression by the adeno-associated virus Rep78 transformation suppressor gene product. Cancer Res 51: 3373–3377

Hermonat PL (1994) Adeno-associated virus inhibits human papillomavirus type 16: a viral interaction implicated in cervical cancer. Cancer Res 54: 2278–2281

Hermonat PL, Muzyczka N (1984) Use of adeno-associated virus as a mammalian DNA cloning vector: transduction of neomycin resistance into mammalian tissue culture cells. Proc Natl Acad Sci USA 81: 6466–6470

Herskowitz I (1987) Functional inactivation of genes by dominant negative mutations. Nature 329: 219–222

Huibregtse JM, Scheffner (1994) Mechanisms of tumor suppressor protein inactivation by the human papillomavirus E6 and E7 oncoproteins. Semin Virol 5: 357–367

James W (1991) Towards gene-inhibition therapy: a review of progress and prospects in the field of antiviral antisense nucleic acids and ribozymes. Antiviral Chem Chemother 2: 191–214

Joshi S, van Brunschot A, Asad S, van der Elst I, Read SE, Bernstein A (1991) Inhibition of human immunodeficiency virus type 1 multiplication by antisense and sense RNA expression. J Virol 65: 5524–5530

Kaniewski WK, Thomas PE (1993) Field testing of virus resistant transgenic plants. Semin Virol 4: 389–396

Kantoff PW, Gillio AP, McKachlin JR, Bordignon C, Eglitis MA, Kernan NA, Moen RC, Kohn DB, Yu S-F, Karson E, Karlsson S, Zwiebel JA, Gilboa E, Blaese RM, Nienhuis A, O'Reilly RJ, Anderson WF (1987) Expression of human adenosine deaminase in nonhuman primates after retrovirus-mediated gene transfer. J Exp Med 166: 219–234

Kaplitt MG, Leone P, Samulski RJ, Xiao X, Pfaff DW, O'Malley KL, During MJ (1994) Long-term gene expression and phenotypic correction using adeno-associated virus vectors in the mammalian brain. Nature Genet 8: 148–153

Kotin RM (1994) Prospects for the use of adeno-associated virus as a vector for human gene therapy. Hum Gene Ther 5: 793–801

Kotin RM, Siniscalco M, Samulski RJ, Zhu X, Hunter L, Laughlin CA, McLaughlin S, Muzyczka N, Rocchi M, Berns KI (1990) Site-specific integration by adeno-associated virus. Proc Natl Acad Sci USA 87: 2211–2215

Kung HJ, Boerkel C, Carter TH (1991) Retroviral mutagenesis of cellular oncogenes: a review with insights into the mechanisms of insertional activation. In: Kung HJ, Vogt PK (eds) Retroviral insertional mutagenesis and oncogene activation. Springer, Berlin Heidelberg New York, pp 1–23 (Current topics in microbiology and immunology, vol 171)

LaFace D, Hermonat P, Wakeland E, Peck A (1988) Gene transfer into hematopoietic progenitor cells mediated by an adeno-associated virus vector. Virology 162: 483–486

Lebkowski JS, McNally MM, Okarma TB, Lerch LB (1988) Adeno-associated virus: a vector system for efficient introduction and integration of DNA into a variety of mammalian cell types. Mol Cell Biol 8: 3988–3996

Lee TC, Sullenger BA, Gallardo HF, Ungers GE, Gilboa E (1992) Overexpression of RRE-derived sequences inhibits HIV-1 replication in CEM cells. New Biol 4: 66–74

Liem SE, Ramezani A, Li X, Joshi S (1993) The development and testing of retroviral vectors expressing trans-dominant mutants of HIV-1 proteins to confer anti-HIV-1 resistance. Hum Gene Ther 4: 625–634

Lin X, Dashti A, Schinazi RF, Tang J (1993) Intracellular diversion of glycoprotein GP160 of human immunodeficiency virus to lysosomes as a strategy of AIDS gene therapy. FASEB J 7: 1070–1080

Lisziewicz J, Sun D, Smythe J, Lusso P, Lori F, Louie A, Markham P, Rossi J, Reitz M, Gallo RC (1993) Inhibition of human immunodeficiency virus type 1 replication by regulated expression of a polymeric Tat activation response RNA decoy as a strategy for gene therapy in AIDS. Proc Natl Acad Sci USA 90: 8000–8004

Liu J, Woffendin C, Yang A, Nabel GJ (1994) Regulated expression of a dominant negative form of Rev improves resistance to HIV replication in T cells. Gene Ther 1: 32–27

Lu D, Chatterjee S, Brar D, Wong KK Jr (1994a) High efficiency in vitro cleavage of transcripts arising from the major transforming genes of human papillomavirus type 16 mediated by ribozymes transcribed from an adeno-associated virus-based vector. Cancer Gene Ther 1: 267–277

Lu D, Wong KK Jr, Podsakoff G, Chatterjee S (1994b) Stable and efficient adeno-associated virus-mediated gene transfer into primitive human marrow-derived hematopoietic cells in long term culture. Blood 84: 360a

Lusso P, Veronese F, Ensoli B, Franchini G, Jemma C, deRocco SE, Kalyanaraman VS, Gallo RC (1990) Expanded HIV-1 cellular tropism by phenotypic mixing with murine endogenous retroviruses. Science 247: 848–852

Maciejewski JP, Weichold FF, Young NS, Cara A, Zella D, Reitz MS Jr, Gallo RC (1995) Intracellular expression of antibody fragments directed against HIV reverse transcriptase prevents HIV infection in vitro. Nature Med 1: 667–673

Malim MH, Freimuth WW, Liu J, Boyle TJ, Lyerly HK, Cullen BR, Nabel GJ (1992) Stable expression of transdominant Rev protein in human T cells inhibits human immunodeficiency virus replication. J Exp Med 176: 1197–1201

Marasco WA, Haseltine WA, Chen SY (1993) Design, intracellular expression, and activity of a human anti-human immunodeficiency virus type1 gp120 single-chain antibody. Proc Natl Acad Sci USA 90: 7889–7893

Mayor HD (1993) Defective parvoviruses may be good for your health! Prog Med Virol 40: 193–205

McLaughlin SK, Collis P, Hermonat PL, Muzyczka N (1988) Adeno-associated virus general transduction vectors: analysis of proviral structures. J Virol 62: 1963–1973

Mendelson E, Grossman Z, Mileguir F, Rechavi G, Carter BJ (1992) Replication of adeno-associated virus type 2 in human lymphocytic cells and interaction with HIV-1. Virology 187: 453–463

Miller AD (1992a) Retroviral vectors. In: Muzyczka N (ed) Viral expression vectors. Springer, Berlin Heidelberg NewYork, pp1–24 (Current topics in microbiology and immunology, vol 158)

Miller AD (1992b) Human gene therapy comes of age. Nature 357: 455–460

Miller DG, Adam MA, Miller AD (1990) Gene transfer by retrovirus vectors occurs only in cells that are actively replicating at the time of infection. Mol Cell Biol 10: 4239–4242

Miroshnichenko OI, Ponomareva TI, Tikchonenko TI (1989) Inhibition of adenovirus 5 replication in COS-1 cells by antisense RNAs against the viral E1a region. Gene 84: 83–89

Morgan RA, Looney DJ, Muenchau DD, Wong-Staal F, Gallo RC, Anderson WF (1990) Retroviral vectors expressing soluble CD4: a potential gene therapy for AIDS. AIDS Res Hum Retroviruses 6: 183–191

Mulligan RC (1993) The basic science of gene therapy. Science 260: 926–932

Muro-Cacho CA, Samulski RJ, Kaplan D (1992) Gene transfer in human lymphocytes using a vector based on adeno-associated virus. J Immunother 11: 231–237

Muzyczka N (1992) Use of AAV as a general transduction vector for mammalian cells. Curr Top Micorb Immunol 158: 97–129

Nabel GJ, Fox BA, Post L, Thompson CB, Woffendin C (1994) A molecular genetic intervention for AIDS–effects of a transdominant negative form of Rev. Hum Gene Ther 5: 79–92

Nakagawa Y, Kakimi K, Ling W, Kubo Y, Higo K, Masuda T, Kuribayashi K, Iwashiro M, Komatz Y, Hirama T, Adachi A, Ishimoto A (1994) Inhibition of murine AIDS (MAIDS), development by the transplantation of bone marrow cells carrying the Fv-4 resistance gene to MAIDS virus-infected mice. J Virol 68: 1438–1441

Natsoulis G, Seshaiah P, Federspiel MJ, Rein A, Hughes SH, Boeke JD (1995) Targeting of a nuclease to murine leukemia virus capsids inhibits viral multiplication. Proc Natl Acad Sci USA 92: 364–368

Oelze I, Rittner K, Sczakiel G (1994) Adeno-associated virus type 2 *rep* gene-mediated inhibition of basal gene expression of human immunodeficiency virus type 1 involves its negative regulatory functions. J Virol 68: 1229–1233

Ojwang JO, Hampel A, Looney DJ, Wong-Staal F, Rappaport J (1992) Inhibition of human immunodeficiency virus type 1 expression by a hairpin ribozyme. Proc Natl Acad Sci USA 89: 10802–10806

Ostrove JM, Duckworth DH, Berns KI (1981) Inhibition of adenovirus-transformed cell oncogenicity by adeno-associated virus. Virology 113: 521–533

Perriman R, Delves A, Gerlach WL (1992) Extended target-site specificity for a hammerhead ribozyme. Gene 113: 157–163

Podsakoff G, Wong KK Jr, Chatterjee S (1994a) Stable and efficient gene transfer into non-dividing cells by adeno-associated virus (AAV)-based vectors. J Virol 68: 5656–5666

Podsakoff G, Shaughnessy EA, Lu D, Wong KK Jr, Chatterjee S (1994b) Long term in vivo reconstitution with murine marrow cells transduced with an adeno-associated virus vector. Blood 84: 256a

Poeschla E, Wong-Staal F (1994) Antiviral and anticancer ribozymes. Curr Opin Oncol 6: 601–606

Ptashne M (1988) How eukaryotic transcriptional activators work. Nature 335: 683–689
Rhodes A, James W (1990) Inhibition of human immunodeficiency virus replication in cell culture by endogenously synthesized antisense RNA. J Gen Virol 71: 1965–1974
Richman DD (1993) HIV drug resistance. Annu Rev Pharmacol Toxicol 33: 149–164
Rittner K, Heilbronn R, Kleinschmidt JA, Sczakiel G (1992) Adeno-associated virus type 2-mediated inhibition of human immunodeficiency virus type 1 (HIV-1) replication: involvement of p78*rep*/*p68* Rep and the HIV-1 long terminal repeat. J Gen Virol 73: 2977–2981
Roberts MR, Qin L, Zhang D, Smith DH, Tran AC, Dull TJ, Groopman JE, Capon DJ, Byrn RA, Finer MH (1994) Targeting of human immunodeficiency virus-infected cells by CD8+ T lymphocytes armed with universal T-cell receptors. Blood 84: 2878–2889
Roe TY, Reynolds TC, Yu G, Brown PO (1993) Integration of murine leukemia virus DNA depends on mitosis. EMBO J 12: 2099–2108
Rommelaere J, Cornelis JJ (1991) Antineoplastic activity of parvoviruses. J Virol Methods 33: 233–251
Salter DW, Crittenden LB (1989) Artificial insertion of a dominant gene for resistance to avian leukosis virus into the germ line of the chicken. Theor Appl Genet 77: 457–461
Sarver N, Cantin EM, Chang PS, Zaia JA, Ladne PA, Stephens DA, Rossi JJ (1990) Ribozymes as potential anti-HIV-1 therapeutic agents. Science 247: 1222–1225
Schlehofer JR (1994) The tumor suppressive properties of adeno-associated viruses. Mutat Res 305: 303–313
Schnittman SM, Fauci AS (1994) Human immunodeficiency virus and acquired immunodeficiency syndrome: an update. Adv Intern Med 39: 305–355
Sczakiel G, Pawlita M (1991) Inhibition of human immunodeficiency virus type 1 replication in human T cells stably expressing antisense RNA. J Virol 65: 468–472
Seif I, de Maeyer E, Riviere I, de Maeyer-Guignard J (1991) Stable antiviral expression in BALB/c 3T3 cells carrying a beta interferon sequence behind a major histocompatibility complex promoter fragment. J Virol 65: 664–671
Shepard AA, Polentino P, DeLuca NA (1990) Trans-dominant inhibition of herpes simplex virus transcriptional regulatory protein ICP4 by heterodimer formation. J Virol 64: 3916–3926
Shillitoe EJ, Kamath P, Chen Z (1994) Papillomaviruses as targets for cancer gene therapy. Cancer Gene Ther 1: 193–204
Shimada T, Fujii H, Maier B, Hayashi S, Mitsuya H, Broder S, Nienhuis AW (1991) Trial of antisense RNA inhibition of HIV replication and gene expression. Antiviral Chem Chemother 2: 133–142
Smythe JA, Sun D, Thomson M, Markham PD, Reitz MS Jr, Gallo RC, Lisziewicz J (1994) A Rev-inducible mutant gag gene stably trasferred into T lymphocytes: an approach to gene therapy against human immunodeficiency virus type 1 infection. Proc Natl Acad Sci USA 91: 3657–3661
Staeheli P (1990) Interferon-induced proteins and the antiviral state. Adv Virus Res 38: 147–200
Sullenger BA, Lee TC, Smith CA, Ungers GE, Gilboa E (1990a) Expression of chimeric tRNA-driven antisense transcripts renders NIH 3T3 cells highly resistant to Moloney murine leukemia virus replication. Mol Cell Biol 10: 6512–6523
Sullenger BA, Gallardo HF, Ungers GE, Gilboa E (1990b) Overexpression of TAR sequences renders cells resistant to human immunodeficiency virus replication. Cell 63: 601–608
Tratschin JD, West MH, Sandbank T, Carter BJ (1984) A human parvovirus, adeno-associated virus, as a eucaryotic vector: transient expression and encapsidation of the procaryotic gene for chloramphenicol acetyltransferase. Mol Cell Biol 4: 2072–2081
Trono D, Feinberg MB, Baltimore D (1989) HIV-1 *gag* mutants can dominantly interfere with the replication of the wild-type virus. Cell 59: 113–120
Vaishnav YN, Wong-Staal F (1991) The biochemistry of AIDS. Annu Rev Biochem 60: 577–630
Van der Krol AR, Mol JNM, Stuitje AR (1988) Modulation of eukaryotic gene expression by complementary RNA or DNA sequences. Biotechniques 6: 958–976
Vieillard V, Lauret E, Rousseau V, De Maeyer E (1994) Blocking of retroviral infection at a step prior to reverse transcription in cells transformed to constitutively express interferon beta. Proc Natl Acad Sci USA 91: 2689–2693
von Ruden T, Gilboa E (1989) Inhibition of human T-cell leukemia virus type 1 replication in primary human T cells that express antisense RNA. J Virol 63: 677–682
Wagner EGH, Simons RW (1994) Antisense RNA control in bacteria, phages, and plasmids. Annu Rev Microbiol 48: 713–742
Walsh CE, Nienhuis AW, Samulski RJ, Brown MG, Miller JL, Young NS, Liu JM (1994a) Phenotypic correction of Fanconi anemia in human hematopoietic cells with a recombinant adeno-associated virus vector. J Clin Invest 94: 1440–1448

Walsh CE, Liu IM, Wang S, Xiao X, Hashmi NJ, Zwerdling T, Agarwal R (1994b) In vivo gene transfer with a novel adeno-associated virus vector to human hematopoietic cells engrafted in SCID-hu mice. Blood 84: 256a

Warner JF, Anderson CG, Laube L, Jolly DJ, Townsend K, Chada S, St Louis D (1991) HIV-specific CTL and antibody responses in mice using retroviral vector-transduced cells. Aids Res Hum Retroviruses 7: 645–655

Watson RJ, Clemens JB (1980) A herpes simplex virus type 1 function continuously required for early and late virus RNA synthesis. Nature 285: 329–330

Weerasinghe M, Liem SE, Asad S, Read SE, Joshi S (1991) Resistance to human immunodeficiency virus type 1 (HIV-1) infection in human CD4+ lymphocyte-derived cell lines conferred by using retroviral vectors expressing an HIV-1 RNA-specific ribozyme. J Virol 65: 5531–5534

Weintraub H, Izant JG, Harland RM (1985) Anti-sense RNA as a molecular tool for genetic analysis. Trends Genet 2: 22–25

Werge TM, Biocca S, Cattaneo A (1990) Intracellular immunization. Cloning and intracellular expression of a monoclonal antibody to the p21ras protein. FEBS Lett 274: 193–198

Werness BA, Levine A, Howley PM (1990) Association of human papilloma virus type 16 and 18 E6 proteins with *p53*. Science. 248: 76–79

Woffendin C, Yang ZY, Udaykumar, Xu L, Yang NS, Sheehy MJ, Nabel GJ (1994) Nonviral and viral delivery of a human immunodeficiency virus protective gene into primary human T cells. Proc Natl Acad Sci USA 91: 11581–11585

Wong KK Jr, Chatterjee S (1992) Controlling herpes simplex virus (HSV) infections: intracellular immunization, the way of the future? Curr Top Micro Immunol 179: 159–174 In: Rouse BT (ed) Herpes Simplex virus. Springer, Berlin Heidelberg New York, pp 159–174 (Current topics in microbiology and immunology, vol 179)

Wong KK Jr, Rose JA, Chatterjee S (1991) Restriction of HSV-1 production in cell lines transduced with an antisense viral vector targeting the ICP4 gene. (Vaccine, vol 91) Cold Spring Harbor Laboratory Press, Cold Spring Harbor, New York, pp 183–189

Yamada O, Yu M, Yee J-K, Kraus G, Looney D, Wong-Staal F (1994) Intracellular immunization of human T cells with a hairpin ribozyme against human immunodeficiency virus type 1. Gene Ther 1: 38–45

Yang Q, Chen F, Trempe JP (1994) Characterization of cell lines that inducibly express the adeno-associated virus Rep proteins. J Virol 68: 4847–4856

Yu M, Ojwang J, Yamada O, Hampel A, Rapapport J, Looney D, Wong-Staal F (1993) A hairpin ribozyme inhibits expression of diverse strains of human immunodeficiency virus type 1. Proc Natl Acad Sci USA 90: 6340–6344

Yu M, Poeschla E, Wong-Staal F (1994) Progress towards gene therapy for HIV infection. Gene Ther 1: 13–26

Zaia JA, Chatterjee S, Wong KK, Elkins D, Taylor NR, Rossi JJ (1992) Status of ribozyme and antisense-based developmental approaches for anti-HIV-1 therapy. Ann N Y Acad Sci 660: 95–105

Zhou C, Bahner IC, Larson GP, Zaia JA, Rossi JJ, Kohn DB (1994) Inhibition of HIV-1 in human T-lymphocytes by retrovirally transduced anti-*tat* and *rev* hammerhead ribozymes. Gene 149: 33–39

Zhou SZ, Cooper S, Kang LY, Ruggieri L, Heimfeld S, Srivastava A, Broxmeyer HE (1994) Adeno-associated virus 2-mediated high efficiency gene transfer into immature and mature subsets of hematopoietic progenitor cells in human umbilical cord blood. J Exp Med 179: 1867–1875

zur Hausen H, de Villiers EM (1994) Human papillomaviruses. Annu Rev Microbiol 48: 427–447

Subject Index

Current Topics in Microbiology and Immunology

Volumes published since 1989 (and still available)

Vol. 178: **Parker, Charles J. (Ed.):** Membrane Defenses Against Attack by Complement and Perforins. 1992. 26 figs. VIII, 188 pp. ISBN 3-540-54653-7

Vol. 179: **Rouse, Barry T. (Ed.):** Herpes Simplex Virus. 1992. 9 figs. X, 180 pp. ISBN 3-540-55066-6

Vol. 180: **Sansonetti, P. J. (Ed.):** Pathogenesis of Shigellosis. 1992. 15 figs. X, 143 pp. ISBN 3-540-55058-5

Vol. 181: **Russell, Stephen W.; Gordon, Siamon (Eds.):** Macrophage Biology and Activation. 1992. 42 figs. IX, 299 pp. ISBN 3-540-55293-6

Vol. 182: **Potter, Michael; Melchers, Fritz (Eds.):** Mechanisms in B-Cell Neoplasia. 1992. 188 figs. XX, 499 pp. ISBN 3-540-55658-3

Vol. 183: **Dimmock, Nigel J.:** Neutralization of Animal Viruses. 1993. 10 figs. VII, 149 pp. ISBN 3-540-56030-0

Vol. 184: **Dunon, Dominique; Mackay, Charles R.; Imhof, Beat A. (Eds.):** Adhesion in Leukocyte Homing and Differentiation. 1993. 37 figs. IX, 260 pp. ISBN 3-540-56756-9

Vol. 185: **Ramig, Robert F. (Ed.):** Rotaviruses. 1994. 37 figs. X, 380 pp. ISBN 3-540-56761-5

Vol. 186: **zur Hausen, Harald (Ed.):** Human Pathogenic Papillomaviruses. 1994. 37 figs. XIII, 274 pp. ISBN 3-540-57193-0

Vol. 187: **Rupprecht, Charles E.; Dietzschold, Bernhard; Koprowski, Hilary (Eds.):** Lyssaviruses. 1994. 50 figs. IX, 352 pp. ISBN 3-540-57194-9

Vol. 188: **Letvin, Norman L.; Desrosiers, Ronald C. (Eds.):** Simian Immunodeficiency Virus. 1994. 37 figs. X, 240 pp. ISBN 3-540-57274-0

Vol. 189: **Oldstone, Michael B. A. (Ed.):** Cytotoxic T-Lymphocytes in Human Viral and Malaria Infections. 1994. 37 figs. IX, 210 pp. ISBN 3-540-57259-7

Vol. 190: **Koprowski, Hilary; Lipkin, W. Ian (Eds.):** Borna Disease. 1995. 33 figs. IX, 134 pp. ISBN 3-540-57388-7

Vol. 191: **ter Meulen, Volker; Billeter, Martin A. (Eds.):** Measles Virus. 1995. 23 figs. IX, 196 pp. ISBN 3-540-57389-5

Vol. 192: **Dangl, Jeffrey L. (Ed.):** Bacterial Pathogenesis of Plants and Animals. 1994. 41 figs. IX, 343 pp. ISBN 3-540-57391-7

Vol. 193: **Chen, Irvin S. Y.; Koprowski, Hilary; Srinivasan, Alagarsamy; Vogt, Peter K. (Eds.):** Transacting Functions of Human Retroviruses. 1995. 49 figs. IX, 240 pp. ISBN 3-540-57901-X

Vol. 194: **Potter, Michael; Melchers, Fritz (Eds.):** Mechanisms in B-cell Neoplasia. 1995. 152 figs. XXV, 458 pp. ISBN 3-540-58447-1

Vol. 195: **Montecucco, Cesare (Ed.):** Clostridial Neurotoxins. 1995. 28 figs. XI., 278 pp. ISBN 3-540-58452-8

Vol. 196: **Koprowski, Hilary; Maeda, Hiroshi (Eds.):** The Role of Nitric Oxide in Physiology and Pathophysiology. 1995. 21 figs. IX, 90 pp. ISBN 3-540-58214-2

Vol. 197: **Meyer, Peter (Ed.):** Gene Silencing in Higher Plants and Related Phenomena in Other Eukaryotes. 1995. 17 figs. IX, 232 pp. ISBN 3-540-58236-3

Vol. 198: **Griffiths, Gillian M.; Tschopp, Jürg (Eds.):** Pathways for Cytolysis. 1995. 45 figs. IX, 224 pp. ISBN 3-540-58725-X

Vol. 199/I: **Doerfler, Walter; Böhm, Petra (Eds.):** The Molecular Repertoire of Adenoviruses I. 1995. 51 figs. XIII, 280 pp. ISBN 3-540-58828-0

Vol. 199/II: **Doerfler, Walter; Böhm, Petra (Eds.):** The Molecular Repertoire of Adenoviruses II. 1995. 36 figs. XIII, 278 pp. ISBN 3-540-58829-9

Vol. 199/III: **Doerfler, Walter; Böhm, Petra (Eds.):** The Molecular Repertoire of Adenoviruses III. 1995. 51 figs. XIII, 310 pp. ISBN 3-540-58987-2

Vol. 200: **Kroemer, Guido; Martinez-A., Carlos (Eds.):** Apoptosis in Immunology. 1995. 14 figs. XI, 242 pp. ISBN 3-540-58756-X

Vol. 201: **Kosco-Vilbois, Marie H. (Ed.):** An Antigen Depository of the Immune System: Follicular Dendritic Cells. 1995. 39 figs. IX, 209 pp. ISBN 3-540-59013-7

Vol. 202: **Oldstone, Michael B. A.; Vitković, Ljubiša (Eds.):** HIV and Dementia. 1995. 40 figs. XIII, 279 pp. ISBN 3-540-59117-6

Vol. 203: **Sarnow, Peter (Ed.):** Cap-Independent Translation. 1995. 31 figs. XI, 183 pp. ISBN 3-540-59121-4

Vol. 204: **Saedler, Heinz; Gierl, Alfons (Eds.):** Transposable Elements. 1995. 42 figs. IX, 234 pp. ISBN 3-540-59342-X

Vol. 205: **Littman, Dan R. (Ed.):** The CD4 Molecule. 1995. 29 figs. XIII, 182 pp. ISBN 3-540-59344-6

Vol. 206: **Chisari, Francis V.; Oldstone, Michael B. A. (Eds.):** Transgenic Models of Human Viral and Immunological Disease. 1995. 53 figs. XI, 345 pp. ISBN 3-540-59341-1

Vol. 207: **Prusiner, Stanley B. (Ed.):** Prions Prions Prions. 1995. 42 figs. VII, 163 pp. ISBN 3-540-59343-8

Vol. 208: **Farnham, Peggy J. (Ed.):** Transcriptional Control of Cell Growth. 1995. 17 figs. IX, 141 pp. ISBN 3-540-60113-9

Vol. 209: **Miller, Virginia L. (Ed.):** Bacterial Invasiveness. 1996. 16 figs. IX, 115 pp. ISBN 3-540-60065-5

Vol. 210: **Potter, Michael; Rose, Noel R. (Eds.):** Immunology of Silicones. 1996. 136 figs. XX, 430 pp. ISBN 3-540-60272-0

Vol. 211: **Wolff, Linda; Perkins, Archibald S. (Eds.):** Molecular Aspects of Myeloid Stem Cell Development. 1996. 98 figs. XIV, 298 pp. ISBN 3-540-60414-6

Vol. 212: **Vainio, Olli; Imhof, Beat A. (Eds.):** Immunology and Developmental Biology of the Chicken. 1996. 43 figs. IX, 281 pp. ISBN 3-540-60585-1

Vol. 213/I: **Günthert, Ursula; Birchmeier, Walter (Eds.):** Attempts to Understand Metastasis Formation I. 1996. 35 figs. XV, 293 pp. ISBN 3-540-60680-7

Vol. 213/II: **Günthert, Ursula; Birchmeier, Walter (Eds.):** Attempts to Understand Metastasis Formation II. 1996. 33 figs. XV, 288 pp. ISBN 3-540-60681-5

Vol. 213/III: **Günthert, Ursula; Schlag, Peter M.; Birchmeier, Walter (Eds.):** Attempts to Understand Metastasis Formation III. 1996. 14 figs. XV, 262 pp. ISBN 3-540-60682-3

Vol. 214: **Kräusslich, Hans-Georg (Ed.):** Morphogenesis and Maturation of Retroviruses. 1996. 34 figs. XI, 344 pp. ISBN 3-540-60928-8

Vol. 215: **Shinnick, Thomas M. (Ed.):** Tuberculosis. 1996. 44 figs. XII, 307 pp. ISBN 3-540-60985-7

Vol. 216: **Rietschel, Ernst Th.; Wagner, Hermann (Eds.):** Pathology of Septic Shock. 1996. 34 figs. X, 321 pp. ISBN 3-540-61026-X

Vol. 217: **Jessberger, Rolf; Lieber, Michael R. (Eds.):** Molecular Analysis of DNA Rearrangements in the Immune System. 1996. 43 figs. IX, 224 pp. ISBN 3-540-61037-5

MIX
Papier aus verantwortungsvollen Quellen
Paper from responsible sources
FSC® C105338
www.fsc.org

If you have any concerns about our products,
you can contact us on
ProductSafety@springernature.com

In case Publisher is established outside the EU,
the EU authorized representative is:
Springer Nature Customer Service Center GmbH
Europaplatz 3, 69115 Heidelberg, Germany

Printed by Libri Plureos GmbH
in Hamburg, Germany